Kent Evenson
942-4216
420-4615

MODERN
RECORDING
TECHNIQUES

HOWARD W. SAMS & COMPANY

AUDIO LIBRARY

Audio IC Op-Amp Applications, 3rd Edition
Walter G. Jung

Audio Production Techniques for Video
David Miles Huber

Electronic Music Circuits
Barry Klein

Handbook for Sound Engineers: The New Audio Cyclopedia
Glen Ballou, Editor

How to Build Speaker Enclosures
Alexis Badmaieff and Don Davis

Introduction to Professional Recording Techniques
Bruce Bartlett   (John Woram Audio Series)

John D. Lenk's Troubleshooting & Repair of Audio Equipment
John D. Lenk

Modern Recording Techniques, 2nd Edition
Robert E. Runstein and David Miles Huber

Musical Applications of Microprocessors, 2nd Edition
Hal Chamberlin

Principles of Digital Audio
Ken C. Pohlmann

Sound System Engineering, 2nd Edition
Don and Carolyn Davis

Stereo TV: The Production of Multi-Dimensional Audio
Roman Olearczuk

*For the retailer nearest you, or to order directly from the publisher, call
800-428-SAMS. In Indiana, Alaska, and Hawaii call 317-298-5699.*

MODERN RECORDING TECHNIQUES
Second Edition

Robert E. Runstein
David Miles Huber

HOWARD W. SAMS & COMPANY

A Division of Macmillan, Inc.
4300 West 62nd Street
Indianapolis, Indiana 46268 USA

©1986 by Howard W. Sams & Co., Inc.,
A publishing subsidiary of Macmillan, Inc.

SECOND EDITION
FOURTH PRINTING — 1988

International Standard Book Number: 0-672-22451-8
Library of Congress Catalog Card Number: 85-61848

Edited by: *Frank N. Speights*
Illustrated by: *T. R. Emrick*

Cover photograph by: *Michael Mendelson*.
Copyright 1985, *Mix Magazine*. Used by permission.

Printed in the United States of America

CONTENTS

PREFACE TO THE SECOND EDITION

Since the first publication of *Modern Recording Techniques* over ten years ago, the world of the multitrack recording studio has undergone many changes. In the early 1970s, we saw the multitrack studio rise in popularity throughout the United States, with

- 16-track studios becoming an industry standard
- the rise and acceptance of the 24-track tape machine (both analog and digital)
- digital mastering to compact disc
- new digital "effects" devices
- a major shift in studio acoustic design
- the marriage of the video and the multitrack studio
- the electronic musical instrument coming of age in both the pop field and the home recording market

The list goes on, even as we speak, with new technology and designs being tested and put on the market. If you get the impression that the recording industry is undergoing a great many changes, you're right. Digital recording and processing techniques are revolutionizing the industry. We have much to look forward to from this industry process that is still in its infancy.

No longer do modern recording techniques apply, however, only to the professional recording engineer. Producers, recording artists, home recordists, and video engineers will all benefit from familiarity with the techniques and the equipment encountered in the recording studio. An understanding of the concepts of sound and studio capabilities, as well as their limitations, will enable such persons to better communicate their ideas to the recording engineer and will also help them in making better recordings.

This book introduces the reader to

- equipment, controls, and problems encountered in the modern recording studio
- multitrack recording studio design and operating techniques
- the central concept of a *transducer* and its place in the recording chain
- the fundamentals of sound, studio acoustics, and basic studio design techniques
- amplifier basics, noise reduction, signal processing, microphone techniques, and the conversion of sound energy into electrical energy
- the setup, operation techniques, and procedures used in the process of *recording*, *overdubbing*, and *mixing*
- the theory of disc recording, cassette duplication, and compact disc production

Modern Recording Techniques has been an effective teaching and reference manual for thousands of perspective and professional engineers, producers, and recording artists. This revised and updated edition will introduce the reader to the many recent advances that have been made in the field of professional recording.

DAVID MILES HUBER

Acknowledgements

I would like to thank the following individuals and companies who have assisted in the preparation of this book by providing photographs and technical information: Dave Talbott, AKG Acoustics Inc.; Aphex Systems, Ltd.; Artist X-Ponent Engineering; Nigel Branwell, Audio Design Calrec, Inc.; Richard Newman, Audio Kenetics; Auratone; David Schwartz, CompuSonics Corp.; Crown International; Harold Cohen, DBX Inc.; Digital Entertainment Corp.; Dolby Laboratories, Inc.; Evantide Inc.; Jack Gilfoy, Gilfoy Sound; Hal Gloff, Gotham Audio Corp.; John Eargle, JBL Inc.; Lexicon Inc.; MCI;

David M. Schwartz, Mix publications; North Shore Marketing; Otari Corporation; Sony Corporation of America; Steve Lawson Productions; Studer-Revox America, Inc.; Dan Kingsbury, Symetrix; Synchronous Technologies; and United Recording Electronics Industries.

Special thanks to Garth Hedin and New World Audio for their cheerful assistance, as well as to all the happy Hoosiers who have helped along the way.

PREFACE TO THE FIRST EDITION

The multitrack recording studio has become the source of almost all pop music recordings released in the United States, thus stirring the interest of many people. While I was Chief Engineer and Technical Director of Intermedia Sound Studios in Boston, the studio was approached by a virtually continuous stream of people who wanted to be trained as recording engineers. Most of them had, as their only qualifications, a love for music and perhaps some musical training or knowledge of hi-fi equipment. All of them wanted to learn how the records they enjoyed were made and to participate in making records as engineers, producers, or recording artists.

Although several books have been written about sound and sound studios, they all seem to emphasize either sound reinforcement, film sound, or radio/tv broadcast sound. Pop music recording has been only mentioned in passing, if at all. The purpose of this book is to fill the information gap for novice and experienced engineers, record producers, and recording artists.

The recording engineer's function is to act as an interface between the producer of a session and the studio equipment. He must translate the producer's ideas into microphone placement and electronic adjustments, and create a high-quality recording of the artist's performance. To do this effectively, the engineer must be well versed in the function, operation, and limitation of each control and piece of equipment in the studio, and he must know how to use them creatively to produce the desired results.

This book introduces the reader to the equipment and controls he will encounter in the modern multitrack recording studio in terms of both the operating techniques currently in use and the roles that they play in creating the finished product. An understanding of the information contained herein, combined with sufficient session observation time for familiarization with the location of the different pieces of equipment in a particular studio, will enable a potential engineer to progress rapidly from observer to assistant engineer. At this point, he will begin operating the equipment and perhaps handling simple sessions on his own. Further experience at the controls during sessions will lead him to full engineer status.

Producers and recording artists will benefit from familiarity with the techniques and equipment they will be using in the studio. Understanding the concepts of sound and studio capabilities and limitations will enable them to better communicate their ideas to the engineer and create better records. More experienced engineers will find several topics and techniques with which they may not be acquainted. Interlocked tape machines have only recently seen much use in pop music recording, and automated mixdown and quad discs are entirely new fields.

The acoustical, mechanical, electrical, and magnetic transformations of a signal from live performance to reproduction from disc are outlined. Then, the physical concepts and terminology used in recording as they pertain to human perception of sound waves and the conversion of sound energy to electrical energy by microphones is discussed.

Next, magnetic tape recording and the function and operation of the components and controls of magnetic tape recorders are detailed while the chapter on signal processing will enable the reader to understand and duplicate many of the special effects heard on current records. The studio equipment, from mike input to tape and monitor speaker outputs, and the operation of each console control is then discussed in detail. The noise-reduction systems described will provide a means of overcoming some of the limitations of present-day magnetic tapes. Monitor speakers, their components, and their interaction with the listening room are other topics that are covered.

As a guide to the novice engineer, the setup, operational techniques, and procedures used in recording, overdubbing, mixing, and sequencing sessions are presented. In addition to the techniques of interlocking the speeds of several tape machines, two different automated mixdown systems currently on the market are introduced and the significance of automation in this area is discussed.

The theory of disc recording is introduced next, and the cutting lathe and mastering console used to transfer signals from tape to disc are detailed. Finally, quadraphonic sound and its storage on a two-channel disc, along with three of the encoding systems vying for dominance in this field, are presented and compared.

ROBERT E. RUNSTEIN

Acknowledgements

I would like to thank the following individuals and companies who have assisted in the preparation of this book by providing photographs and technical information: Allison Buff, Allison Research, Inc.; Charles Overstreet, Altec Corp.; Michael L. Ayers and J. A. Fisher, Ampex Corp.; Lou Lindauer, Don Richter, and Saul Walker, Automated Processes, Inc.; Richard S. Burwen, Burwen Laboratories Inc.; Larry Blakely, DBX, Inc.; Joseph F. Audith, Dictaphone Corp.; Morley Kahn, Dolby Laboratories, Inc.; William S. Sutherland, Electro-Voice, Inc.; Shárdo, Eventide Clockworks, Inc.; Stephen F. Temmer, Gotham Audio Corp.; Dr. Herbert J. Hopkins, Hopkins Sound Technology; James C. Korcuba, International Telecomm, Inc.; Howard E. Krivoy, James B. Lansing Sound, Inc.; G. C. Harned and Thomas M. Hay III, MCI; Andrew Brakhan, North American Philips Corp.; Ed Bedell, Nortronics Company, Inc.; John P. Maloney, Philips Broadcast Equipment Corp.; Eugene R. Shenk, Pulse Techniques, Inc.; Ron Neilson, Quad/Eight Electronics; Horst Ankermann, Sennheiser Electronic Corp.; Edward L. Miller, Spectra Sonics; Florence S. Towers, Superscope, Inc.; P. G. Konold, Shure Brothers, Inc.; R. F. Burnett, Clyde Donaldson, and Robert L. Johnson, 3M Co.; DeWitt Morris, United Recording Electronics Industries; and Paul Ford, Westlake Audio, Inc.

Thanks also to Serge Blinder of Aengus Enterprises and Tom Rabstenek of the Master Cutting Room, Inc. for their help.

Special thanks go to my wife JoAnne for her help in typing the manuscript from my chicken-scratch handwriting while holding down a job of her own, and for putting up with the amount of time I had to spend concentrating on this book rather than on her.

To my love and wife, JoAnne

ROBERT E. RUNSTEIN

This book is dedicated to Philip and Vivian Williams, Jack Gilfoy, John Borwick, Dr. Martin E. Rickey, and Kimon Swartz.

DAVID MILES HUBER

1 INTRODUCTION

The world of the modern multitrack recording studio (Fig. 1-1) is multifaceted. It is a mixture of such fields as music, acoustics, electronics, production, business, and (ask any recording engineer) psychology with each working together to create an end product—the master recording tape. This master tape is then manufactured into a final saleable form—the product—be that product a record, a cassette, or a jingle for television. It usually takes years to successfully learn how to balance this combination of art and recording technology, but, with practice and constant involvement, the skill does develop.

Not only are there many facets to the recording process, but, also, many stages are involved in the production of a product. *Preproduction planning, recording, overdubbing, mixing,* and *product manufacture* are but a few of these stages, as you shall later discover.

From the preceding, we can conclude that there are many facets involved in the recording process and that the recording engineer, producer, and, often times, the musician need to be knowledgeable, flexible, and (let's not forget the most important ingredient) creative.

The Recording Studio

A person new to the recording studio environment will most likely be awestruck by the amount of equipment in the control room—all the microphones, room dividers, and musical instruments. However, there is a definite order to the makeup of a studio, with each piece of equipment serving a definite role in the overall scheme of things. The modern studio has gone through about 50 years of evolution in order to get to its present technology.

The Studio Environment

The studio is a room that is acoustically tuned or optimized for the sole purpose of getting the best sound possible onto a tape while

Fig. 1-1. The db Recording Studio in North Miami, Florida, showing the MCI JH-652 console *(Courtesy Sony/ MCI Corp.).*

using a microphone pickup. It is structurally isolated to keep outside sounds from entering the room and, thus, getting onto the tape, as well as keeping the internal sounds from leaking out and disturbing the surrounding neighborhood.

Studios vary in size, shape, and acoustic design in accordance with the personal tastes of the owners. They can be specially tailored to best record certain styles of music. For example, a studio that records a great deal of rock music might be physically small in size, with highly absorbent walls to allow for high-volume sounds and separation. On the other hand, a studio designed for orchestral film scoring will be much larger by comparison, possibly with high ceilings, to allow for film projection equipment and a greater number of studio musicians.

Over the last two decades, with the advent of the multitrack recorder, studios have generally decreased in size. This is mostly due to the fact that the musicians need not perform together in the studio at the same point in time. *Multitracking* gives the musician the flexibility to put his or her specific musical part onto tape at different times, in different studios, or, possibly, even in another city. This procedure in the recording process is known as *overdubbing*.

The Control Room

The control room (Fig. 1-2) of a recording studio serves two purposes; it is acoustically optimized to act as a critical listening environment, using carefully placed monitor speakers, and it houses the majority of the studio equipment. At the heart of the control room is the *recording console*.

Fig. 1-2. Typical layout of a control room.

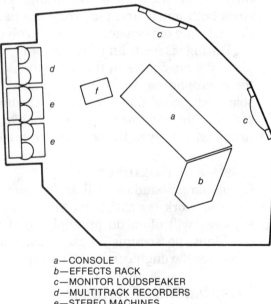

a—CONSOLE
b—EFFECTS RACK
c—MONITOR LOUDSPEAKER
d—MULTITRACK RECORDERS
e—STEREO MACHINES
f—REMOTE CONTROLS FOR RECORDERS

The recording console is the artist's palate of the recording engineer. It allows the mixing together and the control of basically every device found in the studio. The console's basic function is the large task of allowing any combination of inputs, outputs, and effects to be made, which allows all stages of the recording to be quickly and flexibly accomplished.

Tape machines are generally located at the rear of the control room, often with remote controls situated near the engineer for ease of operation. A typical setup will normally involve a multitrack recorder and two stereo machines for mixdown. "Effects" devices, signal processors, and amplifiers will often be placed nearby for accessibility.

Studio Personnel

Before discussing the technology of the recording process, it may be useful to roughly define the roles of the personnel involved in the production of a modern recording session.

The Engineer

The engineer's job can be best described as "an interpreter in a techno-artistic field." It is his job to bring forth the artist's music through the medium of recording. This job is actually an artform since both music and recording are partially subjective, relying upon the taste and experience of those involved.

During a recording session, the studio engineer will generally place the musicians in the desired studio locations, choose and place the microphones, set levels and music balances on the recording console, and record them onto tape. During an overdubbing or mixdown session, the engineer will get balances from the multitrack tape machine, by way of the console, for mixdown or recording purposes.

Assistant Engineer

Often, larger studios will train future staff engineers by allowing them to work (as assistants to engineers) on sessions. The assistant engineer will often do microphone and headphone setups, run tape machines, do session breakdowns, and, in certain cases, get rough mixes for the engineer on the console.

Maintenance Engineer

The maintenance engineer's job is to see that the equipment in the studio is maintained in top condition, regularly aligned, and repaired when necessary. Larger organizations, those with more than one studio, will often employ a full-time maintenance man on the staff.

Many of the smaller studios, however, will be serviced by a free-lance maintenance engineer on an on-call basis, or they may require the engineer to maintain and align his own equipment.

The Producer

The producer, who is most often contracted by the artist or record company on a free-lance basis, serves basically two functions. He will often oversee the music production, giving personal opinions and directions as to how to make the outcome of a recording project the best possible. He may also act as a resource to an artist, who is new to the business, in the areas of project preparation, the session, record pressing, and personal industry contacts. In short, it is his job to oversee a recording project from start to finish.

Transducers

One concept that bears looking into at this point is one that is central to all music, sound, and electronics; this is the concept of the transducer. A *transducer* is any device which changes one form of energy into another corresponding form of energy. The substance which contains the energy is called the *medium*. For example, a violin is a transducer; it will take the vibrations of a bowed string (the medium) and amplify them through a body of wood, converting the vibrations into corresponding sound-pressure waves, which are perceived by us as sound (Fig. 1-3). A microphone is another example. Sound-pressure waves act upon the diaphragm of the microphone and are converted into corresponding electrical voltages. That which is of interest in the recording studio is the manner in which these pressure variations or electrical voltages are stored, so that they can be heard again at a later date.

The electrical impulses from the microphone flow through wires to a series of amplifiers and level controls. They are then fed to a magnetic-tape recorder where they are converted into magnetic impulses and applied to magnetic tape. The tape stores the magnetic impulses in the corresponding order and orientation presented to it, so that the electrical signals can be re-created by the tape recorder *playback head* at a later date. These re-created electrical impulses can then be amplified and fed to a speaker, which converts the impulses to a mechanical motion which, in turn, re-creates the original air-pressure variations that were sensed by the microphone. Alternatively, the electrical impulses from the tape could be fed to the cutting head of a disc mastering lathe, where the impulses would be

**Fig. 1-3. The
violin and
microphone as
transducers.**

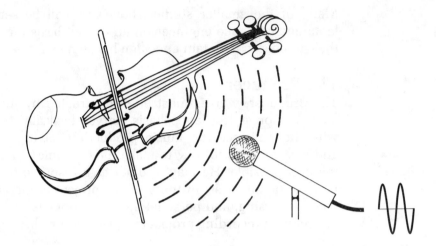

stored in the grooves of a lacquer-covered disc. This disc would then
be used as a mold for the metal plates which press the vinyl discs
found in record stores. The signal is recovered from the vinyl disc
with a phono cartridge while playing the disc. The phono cartridge
converts the impressions found in the grooves of the vinyl disc back
into electrical impulses which can be amplified and fed to a speaker
as before (Fig. 1-4). As can be seen from Table 1-1, transducers can be
found practically everywhere in the audio environment.

**Table 1-1.
Mediums Used
by Transducers
in the Studio To
Transfer Energy**

Transducer	From	To
Ear	Sound waves in air	Nerve impulses in the brain
Microphone	Sound waves in air	Electrical impulses in wires
Record head	Electrical impulses in wire	Magnetic flux on tape
Playback head	Magnetic flux on tape	Electrical impulses in wires
Disc cutter head	Electrical impulses in wires	Grooves cut in disc surface
Phono cartridge	Grooves cut in disc surface	Electrical impulses in wires
Speaker	Electrical impulses in wires	Sound waves in air

Transducers and the mediums they use can often be the weak
link in the chain of an audio system. As we have stated before, a
transducer changes energy in one medium into a corresponding
form of energy in another medium. Given our present technology,
there is no way that this process can be accomplished perfectly.
Noise, distortion, and, often, coloration of the sound are introduced
to a greater or lesser degree. The best that can be done is to minimize
these effects. Differences in design are another factor. Even a slight
design change between one microphone and that of another may
cause the two mikes to sound quite different. This factor, combined

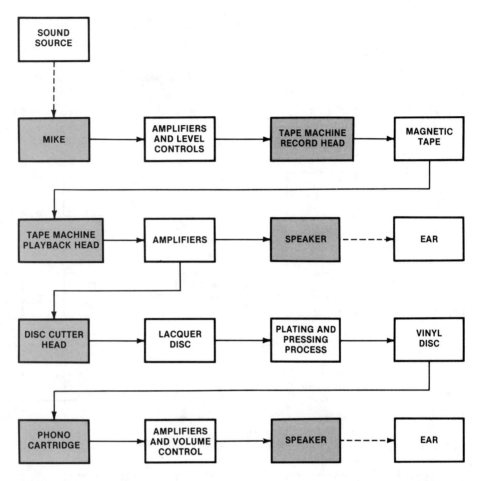

Fig. 1-4. The analog recording chain.

with the complexity of music and acoustics, makes the field of recording a very subjective one.

One device found throughout the recording and audio chains, which is not a transducer, is the amplifier. Both the inputs and outputs of an amplifier are electrical impulses, which allows a greater ease of design and a more reliable and uniform plan of specifications.

Digital recording has the major advantage of causing a vast reduction in the noise and distortion introduced by several of the transducers and mediums used within the recording chain. In a totally digital chain (Fig. 1-5), the acoustic waveforms are picked up by the microphone and converted into electrical impulses. These impulses are then converted into digital form by way of an analog-to-digital (a/d) converter. The a/d converter converts the continuous electrical waveforms into corresponding discrete numeric values, which represent voltage levels. Digital information can be converted back and forth by transducers between the electrical, magnetic, and groove modulation media with virtually no degradation in quality. Thus, if you are listening to a digital recording on a *compact disc (cd)* player at home, the lack of coloration introduced by transducers and mediums in the recording chain, along with the other advantages of digital recording, will give a sonic clarity that is as equally good as listening to the master disc in the studio.

Fig. 1-5. The digital recording chain.

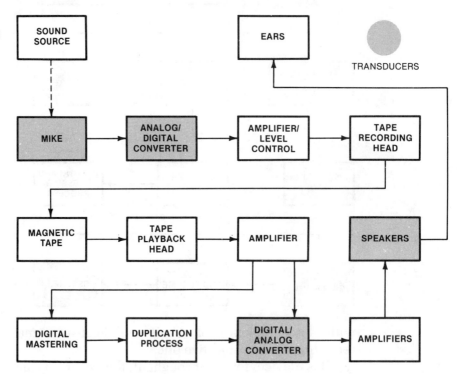

Reference

1. _____, "Playback Workshop Manual," *Playback*, 1985.

2 SOUND AND HEARING

When we make a recording, what we are really interested in doing is capturing and storing sound so that we can either re-create the original sound event at a later date, or use the recorded sound in the creation of a totally new sound event. Starting from the concept that the word *sound* is only a name for the brain's interpretation of a certain type of physical stimulus arriving at the ears, we can divide our examination of sound into three areas: the nature of the stimulus, the characteristics of the ear as a transducer, and the *psychoacoustics* of hearing. This last area, psychoacoustics, deals with how and why the brain interprets a particular stimulus from the ears in just a certain way.

For each sound event in our past experience, our brains have recorded the visual and other sensations that we experienced at the same time. When we are exposed to a new sound event, the brain instinctively recalls the visual and other sensory impressions associated with the most similar sounding event in our past. When we re-create a sound event or create a new one through loudspeakers, however, our current surroundings may conflict with the effect we are trying to generate. For example, if we wish to convey the impression of a sound event occurring in a large hall to a listener in a small room, feeding a close miked signal to the loudspeakers will just not work. The close miked signal does not contain the audio cues necessary to trigger a recall of the listener's impressions of a large hall. In order to make the listener perceive the desired effect, we must alter the close miked signal, making it similar enough to the live sound event that the listener's concentration will focus on past impressions of large halls rather than on his current surroundings. By understanding the physical nature of sound and the affect of the ears in changing sound from a physical phenomenon to a sensory one, we can discover what we need to do to make our recordings convey the desired impressions.

Sound-Pressure Waves

Sound arrives at the ear in the form of a periodic variation in atmospheric pressure, the same atmospheric pressure that is measured by the weatherman with a barometer. The pressure variations corresponding to sound, however, are too small in magnitude and vary too rapidly to be observed on a barometer. These variations in pressure are called sound-pressure waves and can be visualized by imagining the waves seen in a pool of water when a stone is dropped in (Fig. 2-1A). The motion of the water waves moving away from the spot where the stone hit corresponds to the motion of sound-pressure waves moving away from a sound source, except that sound-pressure waves radiate in three directions, not just two.

Sound-pressure waves are generated by a vibrating body that is in contact with the air. This body could be a loudspeaker, someone's vocal cords, or a string from a guitar which vibrates the body of the guitar, which, in turn, vibrates the air next to it, etc. The atmospheric pressure is proportional to the number of air molecules in the area being measured. A vibrating body squeezes additional air molecules into a space as it moves toward the space, creating an area having greater than normal atmospheric pressure, call a *compression*. Since the air molecules were uniformly distributed before motion began, a compression can only be created by borrowing air molecules from the space that the body is moving away from, thereby creating an area with lower than normal atmospheric pressure, called a *rarefaction*. The areas of compression and rarefaction move away from the vibrating body in the form of waves (Fig. 2-1B). The level surface of the water before the stone hits corresponds to the uniform normal atmospheric pressure in the absence of sound, while the hills and valleys present in the water after the stone hits correspond to the levels of atmospheric pressure that are greater and lower than normal, as created by a sound source. The musical pitch of the sound is determined by how often the motion of the body changes direction; this is called the *frequency* of vibration. The loudness of the sound is determined by how far the body moves from the original no-motion position; this is called the *amplitude* of vibration.

Graphs

It is convenient to use graphs when discussing the nature of sound and hearing. A graph is a pictorial representation of the way that one quantity changes with respect to another. For example, if a car is travelling at a speed of 60 miles per hour, we can draw a graph of

Fig. 2-1. Waves created by a stone falling into a pool of water.

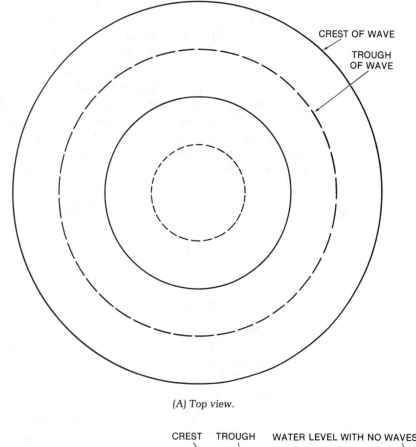

CREST OF WAVE

TROUGH OF WAVE

(A) Top view.

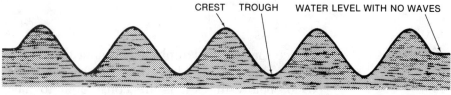

CREST　　TROUGH　　WATER LEVEL WITH NO WAVES

(B) Side view.

how far it has travelled as a function of how long it has been moving. The two variable quantities are elapsed time and distance, since we have fixed the speed at 60 miles per hour. Since the distance travelled depends on how long a time the car has been moving, distance is called the *dependent variable* and time is called the *independent variable*. A graph is drawn by starting with a vertical line and a horizontal line which intersect at a right angle (Fig. 2-2). The horizontal line is called the *X-axis* and is used to represent values of the independent variable (time), while the vertical line is called the *Y-axis* and is used to represent values of the dependent variable (distance). Their intersection point is called the *origin*. Assume that the car is travelling at 60 MPH and that the measurement of time and distance begins at the moment when the car crosses a starting line, like in a race. At the starting line, the time elapsed in hours (t) equals zero, and the distance travelled in miles (d) equals zero. At any time after the car reaches the starting line, we can calculate the distance travelled using the formula d = 60t, so at t = 1 hour, d = 60 miles, and at t = 2 hours, d = 120 miles. If the car was to instantaneously slow down to 30 MPH after two hours had elapsed, it will have travelled a total of 180 miles four hours after it crossed the starting line.

Fig. 2-2. An X-Y graph.

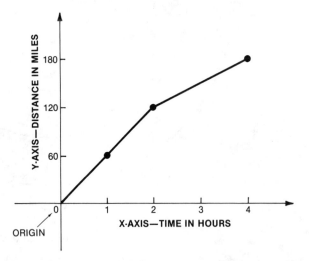

Notice that we have only been using the area of the graph to the right of the Y-axis and above the X-axis. Quantities to the left of the Y-axis or below the X-axis are considered to have negative values; these values do not apply to the distance graph since the car is moving forward. Now, let's look at the graph of a sound-pressure wave (Fig. 2-3). The Y-axis represents the amplitude of the atmospheric pressure with the zero point on the Y-axis corresponding to normal atmospheric pressure. *Compressions*, or increases in pressure, show as positive values above the zero line, while *rarefactions*, or decreases in pressure, show as negative values below the zero line. The X-axis again represents time, but in seconds, not hours, so that the amplitude of the wave can be observed as it changes from moment to moment. By changing the units on the Y-axis, this same graph could represent an electrical waveform or any other type of wave.

Fig. 2-3. A cycle of a wave can be considered to begin at any point on a waveform.

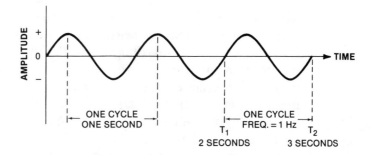

Waveform Characteristics

Every wave has characteristics which distinguish it from other waves; these are frequency, amplitude, velocity, wavelength, phase, harmonic content, and envelope. Let's examine each of these characteristics more fully.

Frequency

In the diagram shown in Fig. 2-3, the value of the waveform starts at zero. At time $t = 0$, the value increases to a maximum in the positive direction, decreases through zero to a maximum in the negative direction, and then returns to zero. Then, the process begins again. One completion of this path is called *one cycle* of the wave. A cycle can begin at any point on the waveform (Fig. 2-4) but, to be complete, it must pass through the zero line and end at a point moving in the same direction (positive or negative) which has the same value as the starting point. Thus, the waveform from $t = 0$ to $t = 2$ constitutes a cycle, and the waveform from $t = 1$ to $t = 3$ is also a cycle. The number of cycles which occur during one second is called the frequency of the waveform and is measured in hertz (Hz). The term *hertz* is equivalent to *cycles per second*.

Fig. 2-4. Graph of waveform amplitude vs. time.

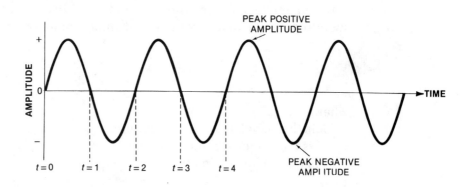

Amplitude

In the case of sound waves, the positive and negative excursions of the curve shown in the diagram of Fig. 2-3 represent increases and decreases in the atmospheric pressure of the air, as it is caused by the sound source and perceived by the listener's ear, with the zero line representing normal atmospheric pressure. The distance above or below the zero line is called the *amplitude* of the waveform at that particular instant of time. The maximum positive and negative excursions are called the *positive* and *negative peak amplitudes*, respectively.

Velocity

The *velocity* of a wave is the speed at which it travels through a *medium* and is given by the equation:

$$V = \frac{d}{t_2 - t_1}$$

(Eq. 2-1)

where,
    V is the wave velocity of propagation in the medium,
    d is the distance from the source,
    t is the time in seconds.

For sound waves, the medium is air molecules; for electricity, the medium is electrons. The wave velocity determines how fast a particular cycle of a waveform will travel a certain distance. At 70 °F, the speed of sound waves in air is approximately 1130 feet per second. This speed is temperature dependent and increases at a rate of 1.1 feet per second for each degree Fahrenheit increase of temperature.

Wavelength

The wavelength (λ) of a wave is the actual distance in the medium between the beginning and the end of a cycle, or between corresponding points on adjacent cycles, and is equal to:

$$\lambda = \frac{V}{f}$$

(Eq. 2-2)

where,
    λ is the wavelength in the medium,
    V is the velocity in the medium,
    f is the frequency in hertz.

To illustrate, a 30-Hz sound wave completes 30 cycles each second, or 1 cycle every ¹⁄₃₀ of a second (approximately every 0.0333 second). The time it takes to complete 1 cycle is called the *period* of the wave and is expressed using the symbol T:

$$T = \frac{1}{f} \qquad \text{(Eq. 2-3)}$$

where,
T is the number of seconds per cycle.

Assuming that sound propagates at the rate of 1130 ft/sec, in the time that it takes 1 cycle to complete, the variation in sound pressure which corresponds to the beginning of that cycle moved 1130 ft/sec for 0.0333 second, and is a distance of 1130 × 0.0333, or 37.66 feet, from the variation in sound pressure which corresponds to the end of that cycle. So, we say that the wavelength of a 30-Hz sound wave in air is 37.66 feet (Fig. 2-5). As the frequency of the waveform increases, each cycle is completed in a shorter amount of time (the period of the wave gets smaller) and the beginning of the waveform cannot travel as far before the end of that cycle is reached. The wavelength, therefore, decreases with increased frequency. For example, ten times as many cycles occur per second in a 300-Hz wave as in a 30-Hz wave, so each cycle of the 300-Hz wave occurs in one tenth the time of a 30-Hz cycle, permitting the beginning of a cycle to travel only one tenth of the distance, or 3.766 feet, before the cycle ends. Since cycles can be measured between any corresponding points on adjacent waveforms, the distance between positive peaks of adjacent waveforms is also one wavelength. The distance between positive and negative peaks of the same cycle would be one half of a wavelength since it is only one half of a cycle. The concept of wavelength shows us that the perception of a wave depends upon the distance from the source as well as time. This distinction will be valuable when we discuss multiple sound sources and the effect of reflections on the perception of sound sources.

Although we speak of wave velocity, the particles of the wave medium do not move far. Sound travels in a compression wave. The air in one spot is compressed by the sound source and this air compresses the air next to it as it returns back to its normal position. The location of the compression moves at the velocity of sound, but the air molecules, which do not travel with the wave, only move to the extent that they are pushed together and, then, are returned to their

Fig. 2-5. Two cyclic wavelengths.

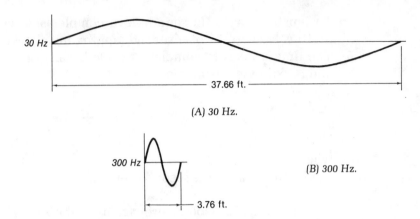

(A) 30 Hz.

(B) 300 Hz.

normal spacing, as in the ball and spring example of Fig. 2-6. This is known as the propagation of a wave. The action of electrical waves in a wire is similar, except that electrons rather than air molecules are propagated.

Fig. 2-6. Propagation of sound waves.

Frequency Response

Now let's take a look at what is called the *frequency response curve* of a device, such as a microphone or an amplifier (Fig. 2-7). In this case, the Y-axis represents the average amplitude of the signal at the output of the device being measured, and the X-axis represents the frequency (or pitch) of the signal. If the input of the device is fed a constant amplitude signal, which rises from the low end to the high end of the scale on the X-axis, the graph will show how the amplitude at the output of the device varies as the frequency of the signal at its input changes. If the output amplitude is the same at all frequencies, the curves will be a flat straight line from left to right. This is where the term *flat frequency response* comes from. It indicates that the device passes all frequencies equally. No frequency is emphasized more or less than the other. If the curve were to dip at certain frequencies, we would know that those frequencies have lower amplitudes than the other frequencies. A frequency-response curve indicates the effect that a device has on the tone of an instrument.

Fig. 2-7. A frequency-response curve.

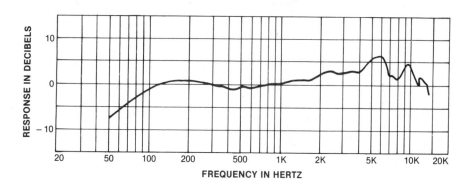

Phase

Since a cycle can begin at any point on a waveform, it is possible to have two wave generators producing sine waves of the same frequency and peak amplitude which will have different amplitudes at any one point in time. These waves are said to be *out of phase* with respect to each other. Phase is measured in degrees (°) and a cycle can be divided into 360°. The sine wave (so named because its amplitude follows the trigonometric sine function) is usually considered to begin at 0° with 0 amplitude, and then increase to a positive maximum at 90°, decrease to zero at 180°, increase again to a maximum but in a negative direction at 270°, and again return to zero at 360°. The first wave (A) in Fig. 2-8 can be considered as our reference curve. The second waveform (B) reaches its maximum positive amplitude 90° before the first and is out of phase with the first because it leads it by 90°. The third waveform (C) begins decreasing from zero 180° before the first and is 180° out of phase. The fourth (D) leads the first by 270° and is also out of phase.

Note that it is assumed that the waves begin at zero degree time; if they had begun earlier, we could have said that wave 1 lagged behind wave 2 by 90°. A wave which is a multiple of 360° out of phase with another can be considered to be in phase with it if the

Fig. 2-8.
Demonstrating
phase
relationships of
sine waves.

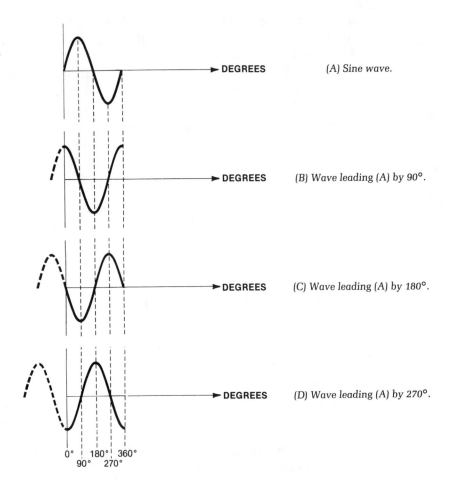

(A) Sine wave.

(B) Wave leading (A) by 90°.

(C) Wave leading (A) by 180°.

(D) Wave leading (A) by 270°.

waves are continuous; there would be no way to distinguish which one of the waves actually lagged behind the other. However, if the waveforms were observed to begin, end, or change in amplitude or frequency, the lagging waveform would do so late and could thus be distinguished.

Waveforms can be added by adding their signed amplitudes at each instant of time. When two waveforms which are completely in phase (0° phase difference) and of the same frequency, shape, and peak amplitude are added, the resulting waveform is of the same frequency, phase, and shape, but will have twice the original peak amplitude (Fig. 2-9A). If two waves are the same as the ones just described, except that they are completely out of phase (phase difference of 180°), they will completely cancel each other when added, resulting in a straight line of zero amplitude (Fig. 2-9B). If the second wave is only partially out of phase (not exactly 180° or (2n − 1) × 180° out of phase), it would *interfere constructively* at points where the amplitudes of the two waves have the same sign (that is, both positive or both negative), resulting in a greater amplitude in the combined wave than in the first wave at that point in time, and it would *interfere destructively* at points where the signs of the two wave amplitudes are opposing, resulting in a lesser amplitude at those points in time than with the first wave (Fig. 2-9C). The waves can be said to be in phase, or *correlated*, at points where the signs are the same and out of phase, or *uncorrelated*, where the signs are opposing.

Phase shift is a term which describes the amount of lead or lag in one wave with respect to another, and is the result of a time delay in the transmission of one of the waves. For example, a 500-Hz wave completes one cycle every 0.002 second. If we start with two in-phase 500-Hz waves and delay one of them by 0.001 second (half the period of the wave), the delayed wave will lag the other by one half a cycle, or 180 degrees. The number of degrees of phase shift introduced by a time delay can be computed by the formula:

$$\phi = \Delta t \times f \times 360°$$

(Eq. 2-4)

where,
    ϕ is the phase shift in degrees,
    Δt is the time delay in seconds,
    f is the frequency in Hz.

From the formula, you can see that the amount of phase shift which results from a fixed time delay varies in direct proportion to the frequency involved. Plugging in the values of a few different frequencies shows that for a 1-millisecond (0.001 second) time delay, we will have the following phase shifts at different frequencies: 250 Hz, 90°; 500 Hz, 180°; 1000 Hz, 360°; 1500 Hz, 540° − 360° = 180°; 2000 Hz, 720° − (2 × 360°) = 0°; 2500 Hz, 900° − (2 × 360°) = 180°; and so forth. Every thousand hertz is a whole-number multiple of 360° and appears in phase with the original frequency, while every one-half thousand hertz is 180°, or a whole-number multiple of 360° plus 180°, and therefore appears out of phase with the original.

If we combine a signal at equal amplitude with the same signal that was delayed by 1 millisecond, the amplitude of the combination

**Fig. 2-9. Adding
sine waves.**

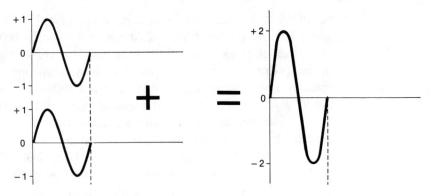

(A) The amplitudes of in-phase waves add when they are mixed.

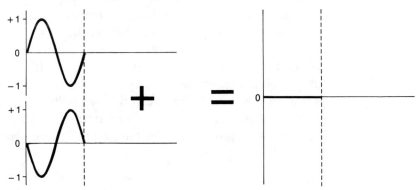

(B) Waves of equal amplitude cancel completely when mixed 180° out of phase.

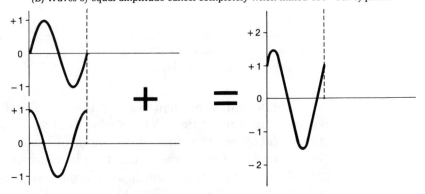

(C) The amplitudes of partially out-of-phase waves add in some places
and subtract in others when mixed.

will increase by a factor of two at the frequencies with 0-degrees
phase shift, and will cancel completely at those frequencies with 180
degrees of phase shift. Frequencies that are phase shifted exactly 90°
will combine with the nondelayed signal with equal amounts of
constructive and destructive interference, resulting in the same

amplitude for the combined wave as for either wave alone. Those frequencies that are shifted between 90° and 180° will have more destructive interference, producing a sum smaller than either signal alone, while those shifted between 0° and 90° will have more constructive interference, producing a sum greater than either signal alone. Except for the 0- and 180-degree cases, the combined signal will be shifted in phase somewhere in between the two original signals.

Any time that a signal follows different paths to the same point, such that the energy from one path is delayed in time with respect to the energy from another path, a frequency-dependent phase difference exists between waves following the two paths. If the energy from the different paths is added together, peaks and dips in the frequency response will be created, for some frequencies will be boosted by constructive interference while others will be lowered in level by destructive interference. The most common source of this type of time delay is distance. For example, if the same source is picked up by two microphones at different distances from the source, a time delay corresponding to the path-length difference will exist. A second source of time delay is in the distance travelled by a reflected sound being picked up by the same microphone that picks up the direct sound. The signals will be in phase at frequencies where the path-length difference is equal to the wavelength of the signals, and out of phase at those frequencies where the path-length difference is one half of the wavelength of the signals. The result of both of these time delays is distorted frequency response. With continuous tones, the interference created by the phase shifts at different frequencies will occur for long delays as well as short ones, but for the majority of sounds in the studio there is a certain time delay beyond which the interference is no longer noticeable due to changes in the signal. At this point, which is longer than about 3 to 5 milliseconds, depending on the character and frequency of the sound, the delayed signal begins to sound like a second source playing in unison with the original. To keep the interference above 20 kHz and, thus, out of the audio range, the path-length difference must be less than 0.34 inch, which corresponds to a time delay of 0.03 millisecond. Since this is such a small amount of delay, you can see that virtually any reflection or time-delayed signal of sufficient level will cause extreme frequency-response degradation. To avoid this, we must either eliminate the reflections or reduce their level to the point where they cannot produce audible cancellations. This is one reason we try to avoid leakage between instruments when we record.

Harmonic Content

Up to this point, we have been discussing the sine wave—which is composed of a single frequency and produces a pure sound at a certain pitch. Musical instruments rarely produce pure sine waves, however, and we are fortunate that they do not. If they did, all instruments playing the same musical note would sound exactly the same and music would be very uninteresting. The factor that enables us to differentiate between instruments is the presence of several different frequencies in the sound wave, in addition to the one that corresponds to the note being played, which is called the *fundamental*. The frequencies present in a sound, other than the fundamental, are called the *partials,* and partials which are higher than the fundamental frequency are called *upper partials* or *overtones*. For most musical instruments, the frequencies of the overtones are whole-number multiples of the fundamental frequency and are called *harmonics* (Fig. 2-10). For example, the frequency corresponding to concert A is 440 Hz. An 880-Hz wave is a harmonic of the 440-Hz wave because it is two times the 440-Hz frequency (Fig. 2-10B). In this case, the 440-Hz wave is called the *fundamental* or *first harmonic* because it is one times the fundamental frequency, and the 880-Hz wave is called the *second harmonic* because it is two times the fundamental. The third harmonic would be three times 440 Hz, or 1320 Hz (Fig. 2-10C). Some instruments, such as bells, xylophones, and other percussive instruments, have partials which are not harmonically related to the fundamental.

The ear perceives those sounds with a frequency ratio of 2:1 to be specially related, and this relationship is the basis of the musical *octave*. For example, since concert A is 440 Hz, the ear hears 880 Hz as having a special relationship to concert A, namely that it is the first tone higher than concert A which sounds most like concert A. The next note above 880 Hz that sounds most like 440 Hz is 1760 Hz. Therefore, 880 Hz is said to be one octave above 440 Hz, and 1760 Hz is said to be two octaves above 440 Hz. Two notes which have the same fundamental frequency and which are played at the same time are said to be in unison, even if they have different harmonics. The human ear does not respond to all frequencies of waves. Its range is limited to the $10\frac{1}{2}$ octaves from about 15 Hz to 20 kHz. Some young people can hear as high as 23 kHz, but the ear's high-frequency response drops off with increased age, and few people over 60 years of age can hear above 8 kHz.

Since the sound waves produced by musical instruments contain harmonics in various amplitude and phase relationships, the wave-

Fig. 2-10.
Illustrating
harmonics.

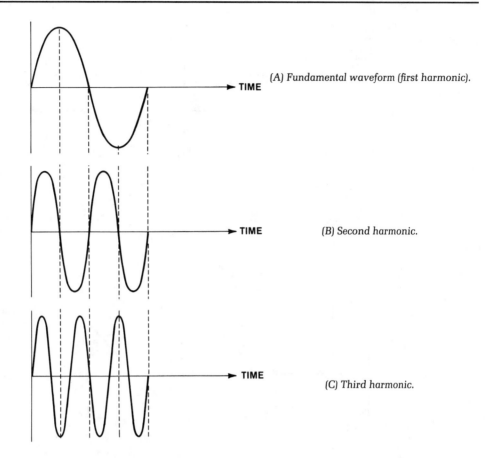

(A) Fundamental waveform (first harmonic).

(B) Second harmonic.

(C) Third harmonic.

forms bear little resemblance to the shape of the single-frequency sine wave. Musical waveforms can be divided into two categories: simple and complex. *Square waves, triangle waves,* and *sawtooth waves* are examples of simple waves containing harmonics (Fig. 2-11). These waveforms are called "simple" because they are continuous and repetitive. One cycle of a square wave looks exactly like the next, and they are all symmetrical about the zero line. The seven wave characteristics mentioned earlier apply to simple waves containing harmonics as well as to sine waves. *Complex waves,* on the other hand, are waves which do not necessarily repeat and which are not necessarily symmetrical about the zero line. An example of a complex waveform is the one created by the speaking of a word (Fig. 2-12). Since complex waves often do not repeat, it is difficult to divide them into cycles or categorize them as to frequency by looking at the waveshape.

Fig. 2-11. Simple waveforms.

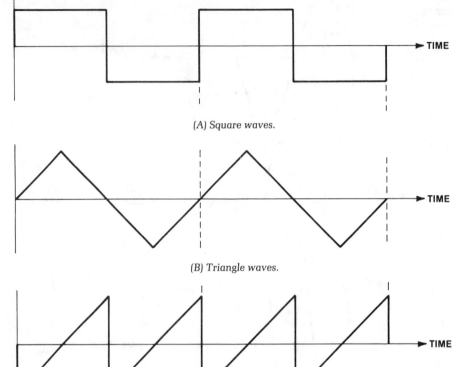

(A) Square waves.

(B) Triangle waves.

(C) Sawtooth waves.

Fig. 2-12. A complex waveform.

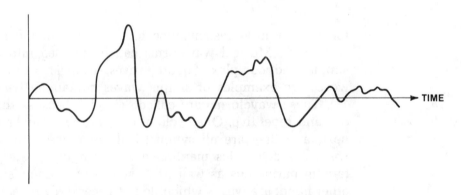

Regardless of the shape or complexity of the waveform reaching the eardrum, the inner ear separates the sound into its component sine waves before transmitting the stimulus to the brain. For this reason, it is not the shape of a waveform that interests us as much as the components that cause it to have that shape, for these components determine the character of sound which the brain perceives. The action of the inner ear can be illustrated by passing a square wave through a bandpass filter which is set to pass only a narrow band of frequencies at any one time. This would show that the square wave is composed of a fundamental frequency plus all of the harmonics whose frequencies are odd-number multiples of the fundamental, with the amplitude of the harmonics decreasing as their frequency increases. In Fig. 2-13, we can see how individual sine-wave harmonics combine together to form a square wave, subtracting from the fundamental where they are uncorrelated and adding to the fundamental where they are correlated.

If we were to analyze the harmonic content of the waves produced by a violin and compare them to the content of the waves produced by a viola, when both are playing concert A (440 Hz), we would obtain the results shown in Fig. 2-14. Notice that the violin has a set of harmonics differing in both degree and intensity from that of the viola. The harmonics present and their relative intensities determine the characteristic sound of each instrument and is called the *timbre* of the instrument. Were we to change the balance of the harmonics, we would change the sound character of the instrument. For example, if the level of violin harmonics 4 through 10 were reduced and the harmonics above the tenth were eliminated, the violin would sound just like the viola.

The significance of harmonics to our perception of tone quality was summarized by Hamm in 1973 as follows:

> The primary color characteristics of an instrument is determined by the strength of the first few harmonics. Each of the lower harmonics produces its own characteristic effect when it is dominant or it can modify the effect of another dominant harmonic if it is prominent. In the simplest classification, the lower harmonics are divided into two tonal groups. The odd harmonics (third and fifth) produce a 'stopped' or 'covered' sound. The even harmonics (second, fourth, and sixth) produce 'choral' or 'singing' sounds. . . . Musically, the second is an octave above the fundamental and is almost inaudible; yet it adds body to the sound, making it fuller. The third is termed a quint or musical twelfth. It produces a sound many musicians refer to as 'blanketed.' Instead of making the tone fuller, a strong third actually makes the tone softer. Adding a fifth to a strong third gives the sound a metallic quality that gets annoying in character as its amplitude increases. A strong second with a strong third tends to open the 'covered' effect. Adding the fourth and fifth to this changes the sound to an 'open horn' like character. . . . The higher harmonics, above the seventh, give the tone 'edge' or 'bite.' Provided the

**Fig. 2-13.
Obtaining a
square wave by
adding odd
harmonics.**

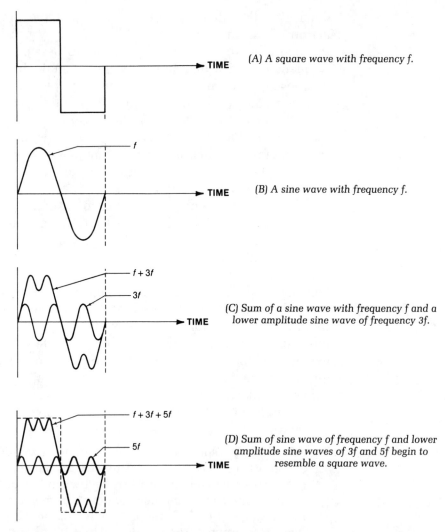

(A) A square wave with frequency f.

(B) A sine wave with frequency f.

(C) Sum of a sine wave with frequency f and a lower amplitude sine wave of frequency 3f.

(D) Sum of sine wave of frequency f and lower amplitude sine waves of 3f and 5f begin to resemble a square wave.

edge is balanced to the basic musical tone, it tends to reinforce the fundamental, giving the sound a sharp attack quality. Many of the edge harmonics are musically unrelated pitches, such as the seventh, ninth, or eleventh. Therefore, too much edge can produce a raspy dissonant quality. Since the ear seems very sensitive to the edge harmonics, controlling their amplitude is of paramount importance. (The study of a trumpet tone) shows that the edge effect is directly related to the loudness of the tone. Playing the same trumpet note loud or soft makes little difference in the amplitude of the fundamental and the lower harmonics. However, harmonics above the sixth increase and decrease in amplitude in almost direct proportion to the loudness. This edge balance is a critically important loudness signal for the human ear.[2]

**Fig. 2-14.
Harmonic
structure of
concert A.**

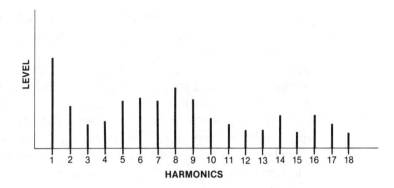

(A) Played on a violin.

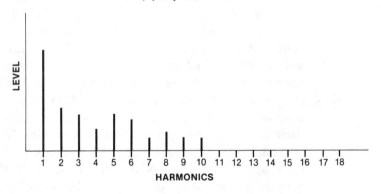

(B) Played on a viola.

Since the relative balance of an instrument's harmonics is so important to its sound, the *frequency response* of microphones, amplifiers, speakers, and all other elements in the signal path can have an affect on the sound. If the frequency response is not flat, the timbre of the sound will be changed. For example, if the high frequencies are amplified less that the low and middle frequencies, the sound will be duller than it should be. *Equalizers* can be used to vary the timbre of instruments, thus changing their subjective affect on the listener. *Phasers* and *flangers* are also timbre modifiers.

Envelopes

Timbre is not the only characteristic that enables us to differentiate between instruments. Every instrument produces its own characteristic *envelope* which works in combination with its timbre to determine the subjective sound of the instrument. The envelope of a waveform describes the way that its intensity varies and can be viewed on a graph by connecting the peak points of the same polarity over a series of cycles. A sound's envelope is composed of three sections: attack, internal dynamics, and decay. *Attack* is the manner in which the sound begins and increases in intensity. *Internal dynamics* describes volume increases, decreases, and sustentations after the attack period, while *decay* is the manner in which the sound ceases. Each of these sections has three variables: time duration, amplitude, and amplitude variation with time. Fig. 2-15A illustrates the envelope of a clarinet note. The attack and decay times are long, and the internal dynamics consist of *sustain*, producing a smooth flowing sound. The next sketch (Fig. 2-15B) illustrates the envelope of a snare drum beat. Note that the initial attack has a much greater amplitude than the internal dynamics, and the attack, initial decay, and final decay are fast, resulting in a sharp crack at the start and a short percussive sound. A cymbal crash (Fig. 2-15C) has a similar high-amplitude fast attack with a fast initial decay, but it sustains longer and decays lower, combining the sharp impulse sound with a smooth lingering shimmer. An organ tone has very rapid attack and decay times and a constant internal amplitude, unless the volume pedal is used to vary the envelope. It can produce sounds varying from a click (if sustain is very short) to smooth sounds if the attack and decay times are made long by the volume pedal. Envelopes with short attack times, followed by fast initial decays, are characterized as sounding percussive or *punchy*, while slow attacks and decays have gentler, smoother sounds. It is important to note that the concept of envelopes uses peak waveform values, while human perception of loudness is proportional to the average wave intensity over a period of time. Thus, high-amplitude portions of the envelope will not make an instrument loud unless the high amplitude is maintained for a long enough period. Short high-amplitude sections contribute to sound character rather than to loudness. Through the use of amplitude controllers, such as compressors, limiters, and expanders, we can modify the sound character of an instrument by changing its envelope without changing the timbre or the sound.

The Ear

A sound source produces sound waves by alternately compressing and rarefying the air between it and the listener. These compressions cause fluctuations of pressure above and below the normal atmospheric pressure. The ear is a sensitive transducer which responds to these pressure variations by way of a series of related processes occurring within the hearing organs, which make up the ear.

**Fig. 2-15.
Waveforms of
various musical
sounds.**

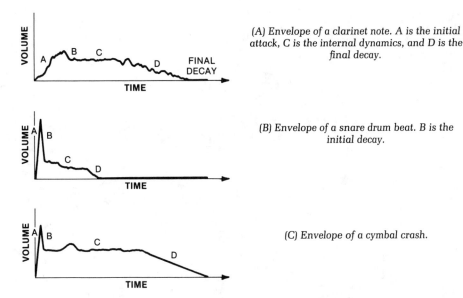

(A) *Envelope of a clarinet note. A is the initial attack, C is the internal dynamics, and D is the final decay.*

(B) *Envelope of a snare drum beat. B is the initial decay.*

(C) *Envelope of a cymbal crash.*

Upon arriving at the listener, sound-pressure waves are collected into the aural canal by way of the outer ear's pinna, and are then directed to the eardrum, a stretched drumlike membrane (Fig. 2-16). The sound waves are then changed into mechanical vibrations and transferred to the inner ear by way of three bones called the *hammer, anvil,* and *stirrup.* These bones act as both an amplifier (by multiplying the vibrations given them by the ear drum many times) and as a limiting protection device (reducing the level of loud transient sounds, such as thunder or fireworks explosions). The vibrations are then applied to the inner ear (cochlea), which is a tubular organ that is rolled up into a snail-like form and which contains two fluid-filled chambers. Within these chambers are tiny hair receptors lined in a row all along the cochlea. Vibrations transmitted to the hairs, which respond to certain frequencies depending upon their placement along the organ, result in the sensation of hearing. Hearing loss generally occurs when these hairs are damaged or deteriorate with age.

Loudness Levels: The dB

The ear operates over an energy range of more than 10^{12}:1 (1,000,000,000,000:1). This is an extremely wide range. Because of its large range, a logarithmic scale has been adopted to compress the measurements of sound into more workable figures. The system used is the *decibel.* The decibel (dB) has no units attached to it; instead, it expresses the ratio of two powers according to the formula dB = 10 log P_1/P_2. The answer is positive if P_1 is greater than P_2 and negative

Fig. 2-16.
Diagram
showing the
outer, middle,
and inner ear.

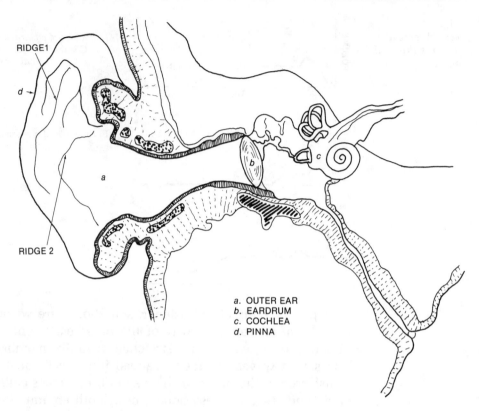

RIDGE1

d

a

b

c

RIDGE 2

a. OUTER EAR
b. EARDRUM
c. COCHLEA
d. PINNA

if P_2 is greater than P_1. If we have two amplifiers such that $P_1 = 60$ watts and $P_2 = 40$ watts, we can express their power ratio in dB as 10 log 60/40 = 10 log 1.5 = 10 times 0.176 = 1.76 dB. We can say that the 60-watt amplifier is 1.76 dB more powerful than the 40-watt amplifier. We could also say that the 40-watt amp is 1.76 dB below that of the 60-watt amp, or that the power of the 40-watt amp is –1.76 dB with respect to the 60-watt amp. Note that we must always state the *reference level* to which the dB value is referred. Since dB represents a ratio, it has no meaning unless the reference level is known. Since the ear responds proportionately to the power level in dB, the 60-watt amp would be 1.76 dB louder than the 40-watt amp if they were alternately connected to the same speaker. As mentioned previously, the ear requires equal ratios of power changes to produce equal loudness increases or decreases. Thus, a 1.76-dB power increase from 40 to 60 watts increases the loudness by the same amount as an increase from 60 to 90 watts because this also represents an increase of 1.76 dB. Under average conditions, the minimum level change that the ear can perceive is about 1.0 dB, although under laboratory or studio conditions of switching sound levels with no appreciable time difference between them, the sensitivity to change can be as low as 0.25 dB at certain frequencies. Thus, an increase from 40 to 60 watts does not bring about much of a loudness increase.

The ear does not hear power directly; the power first must be converted into sound waves. The intensity of the sound waves produced is directly proportional to the power which produced it and, therefore, the ratio of two intensities is equal to the ratio of the two powers which produced them. The ratio of two powers in dB can be computed from their intensity values by plugging the intensity ratio into the dB formula for power, i.e., $dB = 10 \log I_1/I_2$. The intensity of sound is usually measured indirectly, through the measurement of sound-pressure levels (spl). Since the intensity of a sound wave is proportional to the square of the sound pressure, the power in the wave is also proportional to the square of the sound pressure. The ratio of the squares of the two sound pressures is therefore equal to the ratio of the powers that produced them, i.e., $I_1/I_2 = (spl_1)^2/(spl_2)^2 = P_1/P_2$. Power ratios in dB can therefore be expressed by substituting the ratio of the squares of the sound-pressure levels generated into the dB formula for power. Thus, $dB = 10 \log P_1/P_2 = 10 \log (spl_1)^2/(spl_2)^2 = 10 \log (spl_1/spl_2)^2 = 20 \log spl_1/spl_2$. The multiplier in front of the formula changes to 20 for spl ratios because squaring a number doubles the value of its logarithm.

Since the dB system only provides information about the ratio of two powers, it is useful to have a common point, which the number of dB refers to, so that absolute levels can be discussed in terms of dB. For example, if there is a 3-dB difference between two powers, their ratio is two to one. From this information alone, it is impossible to determine whether the power levels in question are 2 watts and 1 watt or 2000 watts and 1000 watts. If, however, a reference level of 1 watt is chosen and a power level is said to be 3 dB above the reference, its level must be twice the reference power, or 2 watts.

Threshold of Hearing

In the case of spl, a convenient pressure-level reference is that of the *threshold of hearing*, which is the minimum sound pressure that produces the phenomenon of hearing in most people, and is equal to 0.0002 microbar. One microbar is equal to one-millionth of normal atmospheric pressure, so it is apparent that the ear is extremely sensitive. In fact, if the ear was any more sensitive, the thermal motion of the molecules in the air would be audible. With the use of 0.0002 microbar as the reference level, spl_2, in the dB formula, is indicated by expressing a value as a certain number of dB spl. The reference pressure level is called 0-dB spl.

The threshold of hearing is defined as the spl for a specific frequency at which the average person can hear only 50% of the time.

Threshold of Feeling

The spl that will cause discomfort in a listener 50% of the time is called the *threshold of feeling* and occurs at a level of about 118-dB spl between 200 Hz and 10 kHz.

Threshold of Pain

The spl which causes pain in a listener 50% of the time is called the *threshold of pain* and corresponds to an spl of 140 dB in the range between 200 Hz and 10 kHz. Fig. 2-17 shows typical spl's for different sounds. The levels are weighted to take into account the reduced sensitivity of human hearing to low frequencies.

The Phon

The unit of loudness level is called the *phon*. Phons are numerically equal to dB spl at 1 kHz and, at other frequencies, are related to dB spl by the Fletcher-Munson curves (Fig. 2-18). Thus, the loudness level of a sound in phons is the number of dB spl produced by a 1-kHz sine wave which sounds equally loud. For example, while 40 phons at one frequency is as loud as 40 phons at any other frequency, 40 phons at 1 kHz requires 40-dB spl, while 40 phons at 10 kHz requires 52-dB spl.

The Sone

Since the perception of the ear to loudness is not directly proportional to the sound-pressure level, a scale has been devised to facilitate the discussion of the loudness of one sound compared to that of another. This unit of loudness is the *sone*, and one sone is defined as the loudness of a 1-kHz sine wave at 40-dB spl. A sine wave is specified because the loudness depends on both the frequency and the complexity of the waveform. A loudness of two sones is twice as loud as one sone and is equal to a level of 50-dB spl at 1 kHz. In the recording studio, the use of dB spl re 0.0002 microbar is more common than phons or sones. The latter are used more by acoustical engineers in designing and examining rooms for specific acoustical properties.

Sound-Pressure Levels

The ear is a *nonlinear* device and, as a result, it produces *harmonic distortion* when subjected to sound waves above a certain loudness.

Fig. 2-17. Typical spl's for common sounds. The reference level of 20 μN/m^2 is equivalent to 0.0002 microbar, or 0.002 dynes/cm^2 (*Courtesy General Radio Co.*).

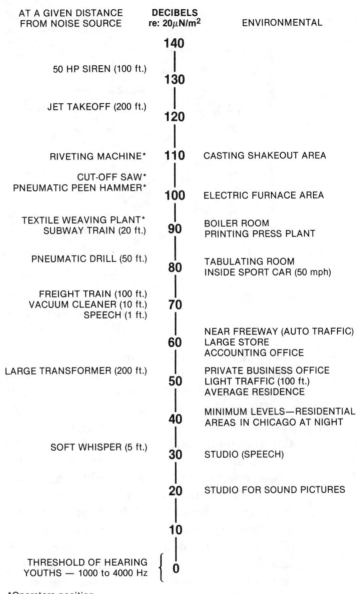

TYPICAL A-WEIGHTED SOUND LEVELS

AT A GIVEN DISTANCE FROM NOISE SOURCE	DECIBELS re: 20μN/m^2	ENVIRONMENTAL
	140	
50 HP SIREN (100 ft.)		
	130	
JET TAKEOFF (200 ft.)		
	120	
RIVETING MACHINE*	**110**	CASTING SHAKEOUT AREA
CUT-OFF SAW* PNEUMATIC PEEN HAMMER*	**100**	ELECTRIC FURNACE AREA
TEXTILE WEAVING PLANT* SUBWAY TRAIN (20 ft.)	**90**	BOILER ROOM PRINTING PRESS PLANT
PNEUMATIC DRILL (50 ft.)	**80**	TABULATING ROOM INSIDE SPORT CAR (50 mph)
FREIGHT TRAIN (100 ft.) VACUUM CLEANER (10 ft.) SPEECH (1 ft.)	**70**	
	60	NEAR FREEWAY (AUTO TRAFFIC) LARGE STORE ACCOUNTING OFFICE
LARGE TRANSFORMER (200 ft.)	**50**	PRIVATE BUSINESS OFFICE LIGHT TRAFFIC (100 ft.) AVERAGE RESIDENCE
	40	MINIMUM LEVELS—RESIDENTIAL AREAS IN CHICAGO AT NIGHT
SOFT WHISPER (5 ft.)	**30**	STUDIO (SPEECH)
	20	STUDIO FOR SOUND PICTURES
	10	
THRESHOLD OF HEARING YOUTHS — 1000 to 4000 Hz	**0**	

*Operators position

Harmonic distortion is the production of waveform harmonics which do not exist in the original signal. Thus, the ear can cause a loud 1-kHz sine wave to be heard as a combination of 1 kHz, 2 kHz, 3 kHz, etc. waves. Although the ear may receive the overtone structure of a violin (if the listening level is loud enough), the ear will produce

Fig. 2-18. The Fletcher-Munson equal-loudness contours.

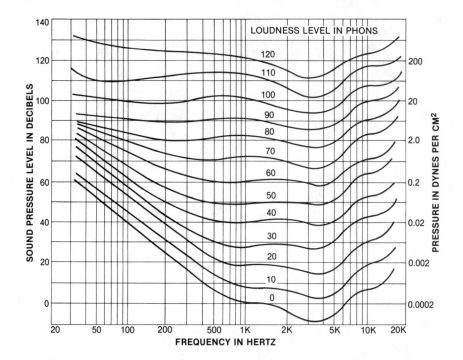

additional harmonics, thus changing the perceived timbre of the instrument. This means that sound monitored at very loud levels may sound quite different when played back at low levels.

The terms *linear* and *nonlinear* are used to describe the output-amplitude versus input-amplitude characteristics of transducers and signal-processing equipment. A linear device or medium is one whose input and output amplitudes have the same ratio at all signal levels. For example, if an amplifier is linear, doubling the input signal amplitude will double the output signal amplitude regardless of the original input signal level. If at certain signal levels, however, doubling the input signal amplitude increases the output signal amplitude by more or less than a factor of two, the amplifier would be called nonlinear at those amplitudes. The use of the term, linear, comes from a graph which is often called the *transfer characteristic* of the device in question. Linear means that the graph is a straight line, while nonlinear means that it either has curves or one or more angles in it. A device can have both linear and nonlinear operating regions. Restricting operation to the linear section avoids distortion, however.

In addition to being nonlinear with respect to amplitude, the ear's frequency response (that is, its perception of timbre) changes with the loudness of the perceived signal. The loudness compensation switch found on many hi-fi preamps is an attempt to compensate for the decrease in the ear's sensitivity to low-frequency sounds at low levels. The curves in Fig. 2-18 are the Fletcher-Munson equal-loudness contours: they indicate the average ear sensitivity to different frequencies at different levels. The horizontal curves indicate the sound-pressure levels that are required to produce the same perceived loudness at different frequencies. Thus, to equal the loudness of a 1.5-kHz tone at a level of 110-dB spl (which is the level typically created by a trumpet-type car horn at a distance of 3 feet), a 40-Hz tone has to be 2 dB greater in sound-pressure level, while a 10-kHz tone must be 8 dB greater than the 1.5-kHz tone to be perceived as being as loud. At 50-dB spl (the noise level present in the average private business office), the level of a 30-Hz tone must be 30 dB greater and a 10-kHz tone must be 14 dB greater than a 1.5-kHz tone to be perceived as being at the same volume. Thus, if a piece of music is monitored so that the signals produce a sound-pressure level of 110 dB, and it sounds well balanced, it will sound both bass and treble deficient when played at a level of 50-dB spl.

From the standpoint that changes in apparent frequency balance (from a playback that is louder or softer than 85-dB spl) is less than for any other monitoring level, 85-dB spl appears to be the best average monitoring level. For example, if the monitoring level is 120-dB spl and we are satisfied with the musical balance, we will note that as we begin to decrease the playback level, the low-frequency response of the ear increases and the bass frequencies appear more pronounced. As the playback level is lowered further, the low-frequency sensitivity of the ear decreases and the bass begins to disappear. Ear sensitivity in the *presence* frequencies also begins to fall. Thus, a mix made at a monitoring level of 120-dB spl will sound bass deficient, distant, and lifeless at lower levels. If we monitor the mix at 100-dB spl and are satisfied with the balance, decreasing the monitoring level will again cause a loss of ear sensitivity, resulting in dips in the presence range, but the greatest change will come in the loss of bass sensitivity where, at 70-dB spl, 50-Hz sounds are heard 12 dB softer than desired. If we monitor at 85-dB spl and are happy with the balance, we can play the mix back at any level between 90- and 60-dB spl and will hear very little change in balance except at the extreme high and low ends of the frequency spectrum where the changes are less than 5 dB. Conveniently, home-listening levels average in the 75- to 85-dB spl range, so 85-dB spl can be considered the optimum monitoring level for mixdowns.

The loudness of a tone can also affect the pitch that the ear perceives. For example, if the intensity of a 100-Hz tone is increased from 40- to 100-dB spl, the ear will perceive a pitch decrease of about 10%. At 500 Hz, the pitch changes about 2% for the same increase in sound-pressure level. This is one reason that musicians find it hard to tune their instruments while listening through headphones. The headphones are often producing higher spls than may be expected.

As a result of the nonlinearity of the ear, tones can interact with each other rather than being perceived separately. Three types of interaction effects occur: *beats, combination tones,* and *masking*.

Beats

Two tones which differ only slightly in frequency and have approximately the same amplitude will produce beats at the ear equal to the difference between the two frequencies. The phenomenon of beats can be used as an aid in tuning instruments because the beats slow down and stop as the two notes approach and reach the same pitch. In a properly tuned piano, not all notes are in perfect tune, and the piano tuner will slightly off tune the instrument by listening to the beat relationships. These beats are the result of the ear's inability to separate closely pitched notes. The resulting synthesis of a third wave represents the addition of the two waves when they are in phase and the subtraction of their intensities when they are out of phase.

Combination Tones

Combination tones result when two loud tones differ by more than 50 Hz, and are the result of *intermodulation distortion* within the ear. The ear will produce an additional set of tones that are equal to both the sum and the difference of the two original tones and which are also equal to the sum and difference of their harmonics. The formulae for computing the tones are:

difference tone frequencies $= af_1 - bf_2$
sum tone frequencies $= af_1 + bf_2$

where a and b are positive integers. The difference tones can be easily heard when they are below the frequency of both the original tones. For example, 2000 and 2500 Hz produce a difference tone of 500 Hz.

Masking

Masking is the phenomenon by which loud signals prevent the ear from hearing softer sounds. The greatest masking effect occurs when the frequency of the sound and the frequency of the masking noise are close to each other. For example, a 4-kHz tone will mask a softer 3.5-kHz tone but will have little effect on the audibility of a quiet 1000-Hz tone. Masking can also be caused by harmonics of the masking tone, so a 1-kHz tone with a strong 2-kHz harmonic could mask a 1900-Hz tone. The masking phenomenon is one of the main reasons that stereo placement and equalization are so important in a mixdown. An instrument which sounds fine by itself can be completely hidden or changed in character by louder instruments with a similar timbre, requiring equalization to make the instruments sound different enough to overcome the masking.

Perception of Direction

One ear cannot discern the direction from which a sound comes, but two ears can. The ability of two ears to localize sound sources in space is called the *binaural effect,* and is the result of the use of three cues received by the ears: (1) interaural arrival time differences, (2) interaural intensity differences, and (3) the effects of the *pinnae,* or the outer ears.

Because the path length to the left ear is longer than that to the right ear, with respect to a sound source on the right, the sound pressure from the source will be sensed later by the left ear than by the right. For a sound source on the extreme left or right, the sound arrives at the far ear with a delay of approximately 620 µsec. For a sound source located 30° off center, the time delay to the far ear would be approximately 240 µsec.

Interaural intensity difference refers to the fact that a sound coming from the right will reach the right ear at a higher intensity level than at the left ear because the left ear is in an *acoustic shadow* cast by the head (Fig. 2-19). The head blocks the direct sound waves and allows only reflected sound from surrounding surfaces to reach the left ear at the blocked frequencies. Since the reflected sound has travelled further and has lost energy at each reflection, the intensity of the sound perceived by the left ear is reduced. This effect is insignificant for low tones because they bend around the head easily, as its diameter is small in comparison to the wavelength of the tones. The effect is considerable for tones above 500 Hz, however, and becomes more significant as the sound-wave frequency rises, causing a difference in the timbre perceived by the two ears. The far ear hears a duller sound.

The interaural delay and intensity cues can tell us the angle from which a sound arrives, but not whether the sound is from in front,

**Fig. 2-19. Sketch
of the head and
source method
of sound
localization.**

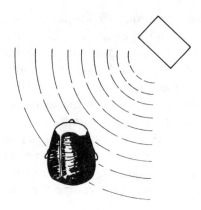

behind, above, or below. The pinnae, however, have two ridges (Fig. 2-16) which produce reflections of the sound that reaches each ear, with the time delay between the sound reaching the entrance of the ear canal directly and the sound reflected from the ridges varying according to the source location. Ridge 1 produces delays of up to 80 μsec which correspond to the location of a source in the horizontal plane. Note that from beyond about 130° from the front there can be no reflection from this ridge because it is blocked by the pinna. So, sounds received with no reflections, which are delayed between 0 and 80 μsec, are perceived as coming from the rear. Ridge 2 produces delays between 100 and 330 μsec which correspond to the location of a source in the vertical plane. The delayed reflections from both ridges combine with the direct sound to produce the characteristic frequency-response colorations that are due to constructive and destructive interference at different frequencies. The brain compares the different colorations at each ear and uses this information to determine the source location. Small movements of the head provide additional information on position due to the changing perspective on the source, but this cue is a minor one compared to the other three.

If there is no difference between what the left and right ears hear, the brain can only assume that the source is at the same distance from each ear. It is this phenomenon which allows the recording engineer to position sound not only in the left and right loudspeakers of a stereo system, but also between the loudspeakers. By feeding the same signal to both loudspeakers, the brain perceives the sound identically in both ears and deduces that the source must be directly in front of the listener. By changing the proportions fed to the two loudspeakers, the engineer changes the interaural intensity difference and can create an illusion that the sound source is anywhere between the two speakers that he desires. The source can even be caused to move between the loudspeakers. This technique is called *panning* (Fig. 2-20) and, although it is the most widely used method, it is not the most effective positioning technique because only those listeners who are equidistant from the left and right loudspeakers will perceive the desired effect. A listener closer to the left loudspeaker will tend to locate the source in that loudspeaker even though it may be panned toward the right. This occurs because the sound from the right loudspeaker must travel further and will arrive slightly delayed, and because the listener's proximity to the left speaker will cause him to receive a higher intensity from that loudspeaker, thereby cancelling the intensity advantage given the right loudspeaker by panning. The other localization cues can also be used by the engineer to assign the source to locations between the two loudspeakers through the use of electronic time delays, phase shifters, and filters.

**Fig. 2-20.
Diagram
showing panpot
vs. spatial
positioning.**

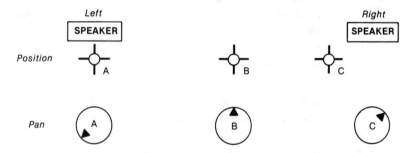

In addition to being able to perceive the direction a sound comes from, the brain can also perceive distance and a sense of the space in which a sound occurs. When sound is generated, it leaves the source in an arc both vertically and horizontally, the angles of which are determined by the nature of the source. Some of the sound reaches the listener without encountering any obstacles and some of it reaches the surrounding surfaces. If these surfaces are hard, they will reflect the sound waves and some of these reflections will also reach the listener. If the surfaces are soft and absorb the waves, or allow the waves to pass through them, little energy will be reflected back to the listener. In air, sound travels at a constant speed of about 1130 feet per second, so the wave that travels a straight line from the source to the listener follows the shortest path and arrives at the listener's ear first. This is called *direct sound*. Those waves that bounce off surrounding surfaces must travel further to reach the listener and, therefore, arrive after the direct sound. These waves form what is called the *reflected sound*, which, in addition to being delayed, can also arrive from different directions than the direct sound. As a result of these additional longer path lengths, the ear hears the sound even after the source stops emitting it. Highly reflective surfaces absorb less of the wave energy at each reflection and enable the sound to persist longer after the source stops than highly absorptive surfaces which dissipate the wave energy. The sound heard in a room can be divided into three successively occurring categories: direct sound, the early reflections, and reverberation. *Direct sound* determines our perception of the sound source location and size, as well as conveying the true timbre of the source. The amount of absorption that occurs when sound is reflected from a surface is not equal at all frequencies. As a result, the timbre of the reflected sound is altered by the characteristics of the surfaces it encountered.

Early reflections reach the ear within 50 milliseconds after the direct sound. These reflections are the result of waves which encountered only a few boundaries before reaching the listener, and they may arrive from a different direction than the direct sound. The time elapsed between hearing the direct sound and the beginning of the early reflections provides information about the size of the performance room, for the further the surfaces are from the listener, the longer it takes the sound to reach them and be reflected back to the listener. The psychoacoustic phenomenon known as the *precedence* or *Haas effect* suppresses our perception of these reflections by as much as 8 to 12 dB, depending on how late after the direct sound they arrive at the ear. The precedence effect applies equally, whether we consider a sound source and its reflection or two spatially separate sound sources, such as two loudspeakers, producing the same sound; the sound will appear to originate at the earlier source even if the delayed source is 8 to 12 dB louder.

Another aspect of the precedence effect is called *temporal fusion*. Early reflections arriving at the listener within 30 milliseconds of the direct sound are not only suppressed in audibility but they are also fused with the direct sound. The ear cannot separately distinguish the closely occurring sounds and considers the reflections to be part of the direct sound. The 30-msec time limit for temporal fusion is not an absolute, rather it depends on the envelope of the sound. Fusion breaks down at 4 milliseconds for transient clicks, while it can extend beyond 80 milliseconds for slowly changing sounds, such as legato violin passages. Despite the fact that the early reflections are suppressed and fused with the direct sound, they do modify our perception of the sound, making it both louder and fuller.

Sounds, reaching the listener more than about 50 milliseconds after the direct sound, have been reflected from so many different surfaces that they begin to reach the listener in a virtually continuous stream and from all directions. These densely spaced reflections are called *reverberation*. Reverberation is characterized by its gradual decrease in amplitude and the warmth and body it gives the sound as well as the loudness it adds to the sound. Since it has undergone multiple reflections, the timbre of the reverberation is quite different from that of the direct sound, with the most notable difference being a roll off of high frequencies and a resultant emphasis of bass. The time it takes for the persisting sound to decrease to 60 dB below its original level is called its *decay time* or *reverberation time* and is abbreviated RT_{60}. The absorption characteristics of a room's surfaces determine its reverberation time. The brain perceives the reverberation time and the timbre of the reverberation and then uses this information to form an opinion on the hardness or softness of the surrounding surfaces. The loudness of the direct sound perceived by the ear increases rapidly as the listener moves closer to the source, while the loudness of the reverberation remains the same because it is so well diffused throughout the room. The perception of the ratio of the loudness of the direct sound to the loudness of the reflected sound enables the listener to judge his distance from the sound source.

To summarize, the direct sound provides information about source location, size, and timbre. The time between perception of the direct sound and the early reflections determines our impression of the size of the performance room. The reverberation decay time provides information about the performance room surfaces, and the proportion of reverberation to direct sound determines our perception of distance from the sound source.

Through the use of artificial reverberation and delay units, the engineer can generate the cues necessary to convince the brain that a sound was recorded in a huge stone-wall cathedral when it was, in fact, recorded in a small dead-sounding room. To do this, the engineer feeds the dry or unreverberated signal into a console fader from which it is split into three paths. One path leads directly to the monitor speakers to provide the direct sound, while the second goes through a time-delay unit and then to the monitor speakers to provide the early reflections. The third path sends the dry signal through a second time-delay unit and then to a reverberation unit to provide the reverberation. The reverberation unit and the first delay unit outputs are connected to additional faders on the console, called *echo return faders,* to control their loudness. Adjusting the number and amount of delays on the first delay unit gives the engineer control of the time that has elapsed between hearing the dry, or direct, sound and the early reflections, thus determining the listener's perception of the size of the room. The second delay unit controls the onset of reverberation; adjusting the reverberation unit's decay time will determine the listener's perception of the room surfaces. A long decay time would indicate a hard-surfaced room, while a short decay time would indicate a soft-surfaced room. By increasing or decreasing the proportion of direct sound to the early reflections and reverberation that is heard by means of the echo return faders, the listener can be fooled into believing the sound source is at the front or rear of the artificial performance room which the engineer has created. Many reverberation units provide a stereo output from a mono input to simulate reverberation coming from directions other than that of the sound source, as occurs in a real room.

Time-delay units can be used independent of reverberation to simulate the dry single-echo effect heard at outdoor live performances, or to simulate additional instruments performing in unison. This is possible because our perception that more than one instrument is playing depends on the lack of exact synchronization between the sounds. No matter how good the musicians are, the fact that each instrument is a different distance away from the listener, no matter how slight, ensures a lack of synchronization due to the different amounts of time required for the waves to reach the listener's ear. By repeating a signal after a short time delay of 5 milliseconds or so, the apparent number of instruments being played is doubled. This process is called *electronic doubling* or *electronic double tracking.* Often, doubling and tripling is done by the artist, himself, through multiple overdubbing, so that one violin can sound like an ensemble, or so that two vocalists sound like a chorus. Doubling can also be used to strengthen weak vocal performances because the slight irregularities in one performance are hidden by the other. If the delay time is long enough so that the repeat is heard discreetly, or more than about 35 milliseconds, the repeat is often called *slap echo,* or *slap back,* and has the effect of causing the rhythm to bounce or double.

Waveform Values

Since waves are constantly varying in amplitude, how can they be said to have a single value, such as a level of 110-dB spl? The maximum value cannot be used because it is reached only for an instant on each side of the zero line during a cycle. In addition, waveforms of different shapes with the same peak amplitudes have different energy content and would not produce the same loudness. Taking the average over a number of complete cycles for any symmetrical wave would result in a value of zero, since the wave would have equal positive and negative excursions. The positive and negative maximums of a wave are called the *peak values*. The difference between the positive and negative peaks is called the *peak-to-peak value*, and the value of the amplitude at any one instant in time is called the *instantaneous value*.

To get a meaningful average of all these values, the *rms* or *root-means-square* value was developed. For a sine wave, this value is arrived at by squaring the amplitude of the wave at each point on the waveform, which brings the waveform above the zero line (thus, it has a nonzero average) since the square of a negative number is a positive number, and the square of a positive number is also a positive number. The average of the squared waveform is taken by dividing the peak amplitude by two, and the rms value is calculated by taking the square root of the result. Beginning, for example, with a peak amplitude of 3 units, the rms value of this wave is the square root of the quantity (3 squared divided by two) or 0.707 times 3. Thus, the rms value of a sine wave is 0.707 times its peak amplitude. The computation of rms values of nonsinusoidal waves is much more difficult requiring that the rms value of each frequency component be computed and summed. All rms values have in common the fact that the rms value of a wave is equal to the value of continuous (i.e., direct current) signal that will produce the same amount of power as the wave in question (Fig. 2-21). For this reason, rms values are also referred to as *continuous* or *effective* values. Unless another value is specified when speaking quantitatively about a wave, this book refers to the rms value.

Fig. 2-21. The level of direct current produces the same amount of power as the sine wave.

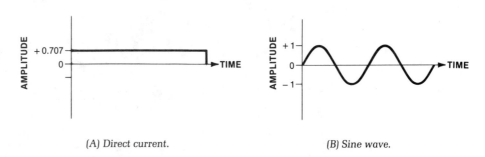

(A) Direct current.   (B) Sine wave.

References

1. Backus, John, *The Acoustical Foundations of Music*, New York, W. W. Norton & Co. Inc., 1969.
2. Hamm, Russell O., "Tubes Versus Transistors—Is There An Audible Difference?" *Journal of the Audio Engineering Society*, Vol. 21, No. 4, May 1973, pp. 267-273.
3. Nisbett, Alec, *The Technique of the Sound Studio*, New York: Hastings House Publishers, Inc., 1972.
4. Peterson, Arnold P. G., and Ervin E. Gross, Jr., *Handbook of Noise Measurement*, 7th Edition, General Radio Co., 1972.
5. Tremaine, Howard M., *The Audio Cyclopedia*, Indianapolis: Howard W. Sams & Co., Inc., 1978.
6. Weidner, Richard T., and Robert L. Sells, *Elementary Classical Physics*, Vol. 2, Boston: Allyn and Bacon, Inc., 1965.

3 STUDIO ACOUSTICS

The *Audio Cyclopedia*, written by Howard M. Tremaine, defines the term *acoustics* as " . . . a science dealing with the production, effects, and transmission of sound waves; the transmission of sound waves through various mediums, including reflection, refraction, defraction, absorption, and interference; the characteristics of auditorium, theaters, and studios, as well as their design."

It is easy to tell from this description that the proper acoustic design of an environment such as a modern recording studio is no simple matter. Many complex variables and interrelations come into play in the making of a successful studio design and they must take into consideration certain basic requirements which, once met, will offer effective results:

1. Acoustic isolation from external noises which are normally transmitted into the studio environment from outside, as well as preventing interior sounds from the getting out.
2. The frequency components of a complex waveform should maintain their relative intensities.
3. The acoustical environment must offer sufficient flexibility and control to allow for optimum intelligibility, as well as allowing acoustic separation within the studio.

Isolation

It is important for studio design to take into account effective isolation techniques in order to reduce external noise, whether it is transmitted through the medium of air or solids (such as the building structure). In order to reduce such extraneous noises as auto or jet traffic, specialized construction techniques well beyond the average may be necessary.

For those who are afforded the luxury of building a studio facility from the ground up, the choice of studio location will often take into heavy consideration such things as neighborhood noise levels, as extensive isolation construction techniques can be very expensive. If there is no choice in the location of the site, and the studio happens to be located under the flight pattern of Delta 3 or over commuter train Number Five, you just have to give in to destiny and use your better judgement. In this situation, you will have to design and build an acoustical barrier to outside interferences. The reduction in the sound pressure level of a sound source through an acoustic barrier (Fig. 3-1) is termed the *transmission loss* of a signal. Transmission loss may be defined as the ratio of the sound pressure intensity incident upon an acoustical barrier (a), to the intensity of the sound transmitted by the barrier (b). The transmission loss, in dB, introduced by the barrier is given by Equation 3-1.

Fig. 3-1. The four factors of a sound wave (incidence, reflection, absorption, and transmission) which affect the pressure level of a sound source passing through an acoustic barrier.

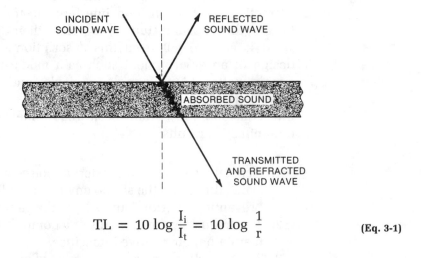

$$ \text{TL} = 10 \log \frac{I_i}{I_t} = 10 \log \frac{1}{r} \qquad \text{(Eq. 3-1)} $$

where,
    TL is the transmission loss,
    I_i equals the intensity of the incident sound,
    I_t equals the intensity of the transmitted sound,
    r is the transmission coefficient.

The transmission coefficients of various construction materials are given in Table 3-1. By obtaining a high transmission loss and, thus, a high degree of isolation, it is possible to build a thick, solid, high-density barrier to achieve the required goal but, in most situations, this would prove to be impractical as well as expensive.

Table 3-1. Absorption Coefficients of Various Acoustical Materials, Building Materials, and Objects

Material	Thickness, inches	Mounting**	Frequency						Author*
			128	256	512	1,024	2,048	4,096	
			Coefficient						
Corkoustic B-5.........	1½	2	0.18	0.41	0.70	0.51	0.58	0.65	A.M.A.
Cushiontone A-3........	⅞	2	0.17	0.51	0.73	0.95	0.75	0.72	A.M.A.
Sanacoustic pad, with metal facing..........	1⅛	3	0.25	0.56	0.99	0.99	0.91	0.82	A.M.A.
Fibretex	13⁄16	2	0.16	0.49	0.56	0.78	0.84	0.78	A.M.A.
Acoustex 4OR.........	¾	2	0.09	0.17	0.59	0.90	0.75	0.73	A.M.A.
Fiberglas tile Type A....	1	2	0.17	0.44	0.91	0.99	0.82	0.77	A.M.A.
Acoustone F	13⁄16	1	0.12	0.31	0.85	0.88	0.75	0.75	A.M.A.
Acousti-Celotex C-4.....	1¼	2	0.25	0.58	0.99	0.75	0.58	0.50	A.M.A.
Draperies hung straight, in contact with wall, cotton fabric, 10 ounces per square yard.................	...	..	0.04	0.05	0.11	0.18	0.30	0.44	P.S.
Same as above, but velour, 18 ounces per square yard	...	..	0.05	0.12	0.35	0.45	0.40	0.44	P.S.
Same as above, hung 4 inches from wall......	...	..	0.09	0.33	0.45	0.52	0.50	0.44	P.S.
Felt, all hair, contact with wall	...	..	0.13	0.41	0.56	0.69	0.65	0.49	P.S.
Rock wool	1	..	0.35	0.49	0.63	0.80	0.83	...	V.K.
Carpet, on concrete	0.4	..	0.09	0.08	0.21	0.26	0.27	0.37	B.R.
Carpet, on ⅛-inch felt, on concrete	0.4	..	0.11	0.14	0.37	0.43	0.27	0.27	B.R.
Concrete, unpainted	...	..	0.010	0.012	0.016	0.019	0.023	0.035	V.K.
Wood sheeting, pine	0.8	..	0.10	0.11	0.10	0.08	0.08	0.11	W.S.
Brick wall, painted......	...	..	0.012	0.013	0.017	0.020	0.023	0.025	W.S.
Plaster, lime on wood lath on wood studs, rough finish	½	..	0.039	0.056	0.061	0.089	0.054	0.070	P.S.
Individual object			**Absorption units, square foot (sabins)**						
Audience, per person, man with coat........	...	..	2.3	3.2	4.8	6.2	7.6	7.0	B.S.
Auditorium chairs, solid seat and back.........	...	..	0.15	0.22	0.25	0.28	0.50	...	P.S.
Auditorium chairs, upholstered	...	..	...	3.1	3.0	3.2	3.4	...	F.W.

*Abbreviations in the above table are as follows: A.M.A., Acoustical Materials Association; W.S., Wallace Sabine; P.S., P.E. Sabine; F.W., F.R. Watson; V.K., V.O. Knudsen; B.R., Building Research Station, England; B.S., Bureau of Standards.

**Mountings in the above table are as follows:
1. Cemented to plaster board.
2. Nailed to 1- by 2-inch furring 12 inches apart.
3. Attached to metal supports applied to 1- by 2-inch wood furring.
4. Laid on 24-gauge sheet iron, nailed to 1- by 2-inch wood furring 24 inches apart.

Courtesy Dover Publications, Inc.

Alternative building techniques have proved to be very effective in maximizing TL with a minimum of thickness, weight, and financial outlay. One major method which falls into the category of providing excellent acoustic isolation is the employment of floating wall and floor construction. This technique of *floating a room* refers to double barrier construction in which the inner barrier is physically separated from the outer barrier either by means of an elastic or isolating support, such as spring supports or rubber absorber construction, or by totally isolating the barriers from each other (Fig. 3-2).

Fig. 3-2. Isolating the inner and outer walls of a room from each other.

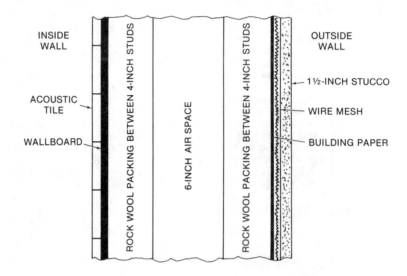

Air space, when placed between two barriers, is found to be an extremely efficient absorption material for providing isolation. The resonant frequencies of the two barrier walls must be carefully controlled such that the TL will remain high over the entire audio spectrum and will not break down between the barriers at some common resonant frequency. In the case of a floated floor, an elastic support, such as springs, provides an excellent means of isolation against any subsonic structure-born frequencies and oscillations which may be transmitted through the building to the studio floor.

Not only is a high degree of isolation required between the studio's interior and exterior environment, an equal amount of isolation is required between the studio and the control room. This is important in order to hear an accurate tonal balance over the control-room monitors, without leakage from the studio coloring the audible signal.

A control room wall is constructed in a manner similar to that used in providing isolation in the studio outer walls, with the exception of the *control-room windows* and *traffic doors* which provide access to and from the studio. Control-room window construction, when properly done, utilizes double-window construction (Fig. 3-3), for the same reason and purpose as does the double-wall construction. This places air space between two panes of glass for a further increase in transmission loss.

Fig. 3-3. Cross-section of control-room window construction.

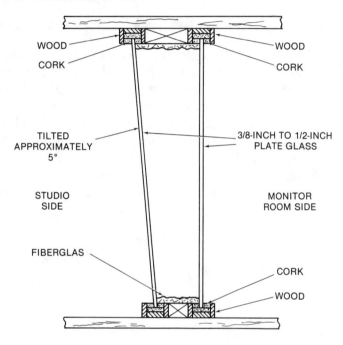

WOOD

CORK

WOOD

CORK

TILTED APPROXIMATELY 5°

3/8-INCH TO 1/2-INCH PLATE GLASS

STUDIO SIDE

MONITOR ROOM SIDE

FIBERGLAS

CORK

WOOD

The glass panels used in window construction are in the ⅜-inch to ½-inch thickness range and are seated in an elastic damping seal, such as rubber, to prevent structure-born oscillations from vibrating the glass. It is important that at least one or both of these panels are tilted at a 5° angle (minimum), with respect to each other, in order to eliminate any possibility of a buildup of standing waves within the sandwiched air space (which would break down the transmission loss at specific frequency intervals).

Access doors to and from the studio, control room, and exterior may also be constructed in a double-door fashion, forming what is known as a *sound lock* (Fig. 3-4). A sound lock, plus the use of floating-room techniques, substantial and double-door construction, and close tolerances, will provide a good seal against any outside acoustic leakage.

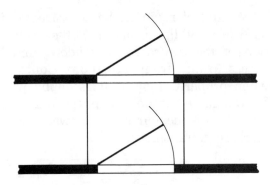

Isolation rooms are acoustic barrier areas that are built into or accessed from the studio and which provide improved separation between loud and soft instrumentalists or sections. These barrier areas may be constructed in several fashions depending on the studio design and usage. An *isolation,* or *iso, booth* may take the form of a totally separate room, with both visual and physical access onto the main studio floor (Fig. 3-5). Construction of this room would most likely be of similar material and design to that used in the control room and studio walls, thus providing maximum isolation.

**Fig. 3-5. An
isolation room
located adjacent
to the studio
area.**

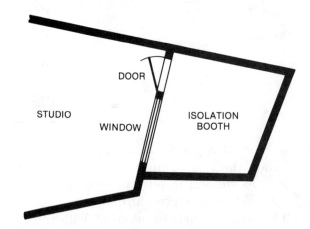

In a trend to improve visual contact and to visually enlarge the studio's atmosphere, movable glass partitions are beginning to be used in modern designs. This method makes use of high-grade, glass, sliding doors similar to those used to access an outdoor patio (Fig. 3-6). This design allows sections of a studio to either be partitioned off (providing isolation) or opened up (to make full use of the studio space) as required by the situation. This allows a great deal of flexibility in studio application. Depending upon the degree of isolation required, the sliding glass door may be of a single or double-barrier construction, with double-barrier construction increasing the transmission loss through the construction of a sound lock.

Fig. 3-6. Sliding-door partitioning.

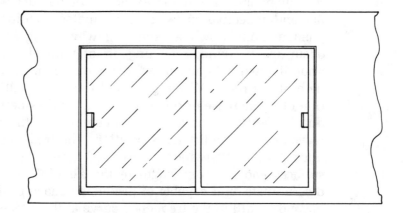

Movable partitions (Fig. 3-7), or *flats* as they are called, may be used to provide an isolation barrier. These flats provide the least amount of isolation within the studio but they allow the greatest degree of flexibility, permitting any and all isolation changes required by most situations. Under most conditions, an adequate degree of isolation can be attained through careful microphone, musician, and flat placement.

Fig. 3-7. Acoustic partition flats.

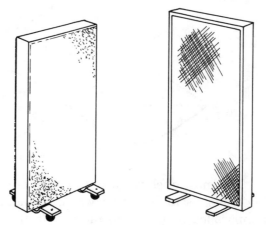

Frequency Balance

The second prerequisite for a properly designed room is that the frequency components of a complex sound wave maintain their relative level intensities. Another way of stating this is that the room should exhibit a relatively flat frequency response over the entire audio range without adding its own colorations to the sound.

In the design of an acoustic environment, the first major rule to remember is to avoid the existence of parallel wall structures with flat and reflective surfaces. This invites conditions conducive to a phenomenon known as standing waves (Fig. 3-8). *Standing waves* occur when sound, reflecting off of parallel surfaces, travels back upon its own path and interferes with the amplitude along the sound path of the room. Walking around the room will produce the sensation of an increase or decrease in the sound's perceived level. This is due to cancellations and reinforcements of the combined reflected waveforms at the listener's position. The frequencies which will produce standing waves are determined by the *distance between parallel surfaces* and the *wavelength of the signal*. This makes it frequency discriminate, and capable of producing sharp peaks or dips (up to or beyond 20 dB) in the frequency curve at the affected fundamental frequency(s), and harmonic intervals. This condition not only exists for parallel walls, but for any parallel surface, such as between floor and ceiling (which can be particularly problematic) or between two reflective flats.

Fig. 3-8.
Examples of
standing waves.

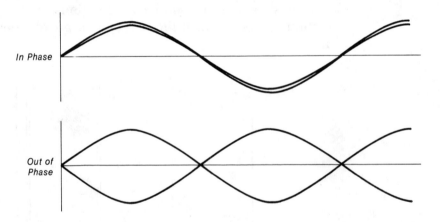

The most effective way to ensure against this condition of standing waves is to construct walls, boundaries, and ceilings which are nonparallel, or break up these reflective boundaries by constructing diffusers (Fig. 3-9). *Diffusers* are acoustical boundaries which reflect the sound wave back at a wider diffused angle than what was received, thereby breaking up the energy-destructive standing-wave condition. The employment of both nonparallel and diffusion wall construction will, in addition, reduce a condition known as *flutter*

echo and will smooth out the reverberation characteristics of the room through increased acoustical pathways.

Fig. 3-9. Two examples of diffusers.

Flutter, or slap, echo is a condition which occurs where parallel boundaries are spaced far enough apart for the listener to discern many discrete echoes. This will often give a smaller room a tube-like hollow sound which affects the sound character as well as the frequency response. A larger room, which may contain delayed echo paths of one second or more, will have its echoes spaced so far apart in time that the reflected sound actually interferes with the intelligibility of the direct sound, resulting in a jumble of noise.

Absorption

Another method of designing a recording studio or control room for a smooth flat response, throughout the acoustic space, is to measure the frequency response of the room with a real-time spectrum analyzer, once the approximate acoustical calculations and construction have been made. This will immediately determine how the room reacts over the entire audio spectrum. Should the room show irregularities in frequency response, even after careful attention to reflections has been paid, the next step is to tailor the frequency response through absorption of those frequencies which are presenting problems.

Just as a mirror reflects light, a smooth rigid acoustical surface will reflect sound from its surface, giving off an amount of energy that is almost equal to what was received at its surface (Fig. 3-10). Absorption of acoustic energy is the inverse of reflection (Fig. 3-11). When sound strikes a material, and a portion of the energy is absorbed and a portion is reflected, it may be expressed as a ratio. This ratio is termed as the *absorption coefficient* of the absorbative material. For a given material, the absorption coefficient, a, is:

$$a = \frac{I_a}{I_r} \qquad \text{(Eq. 3-2)}$$

where,

    I_a is the absorbed sound,

    I_r is the incident sound.

The factor $(1 - a)$ is the reflected sound. This makes the coefficient, a, some figure between 0 and 1. A sample listing of these coefficients is provided in Table 3-2.

Fig. 3-10. Sound reflection.

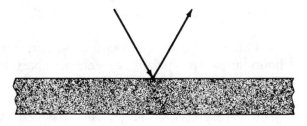

Fig. 3-11. Sound absorption.

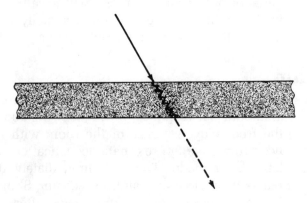

If we say that a surface material has an absorption coefficient of 0.25, we are saying that the material will absorb 25% of the incident acoustic energy, while reflecting back 75% of the total sound energy at that frequency.

High-Frequency Absorption

The absorption of high frequencies (Fig. 3-12) is accomplished through the use of dense porous materials, such as cloth, fiberglas,

**Table 3-2.
Absorption
Coefficients for
Different
Materials**

		Coefficients					
Material	125 Hz	250 Hz	500 Hz	1000 Hz	2000 Hz	4000 Hz
Brick, unglazed..................................	0.03	0.03	0.03	0.04	0.05	0.07
Carpet, heavy on concrete	0.02	0.06	0.14	0.37	0.60	0.65
Carpet, with latex backing on 40-oz hairfelt of foam rubber..	0.08	0.27	0.39	0.34	0.48	0.63
Concrete block, coarse	0.36	0.44	0.31	0.29	0.39	0.25
Light velour, 10 oz per sq-yd in contact with wall ...	0.03	0.04	0.11	0.17	0.24	0.35
Concrete or terrazo	0.01	0.01	0.015	0.02	0.02	0.02
Wood..	0.15	0.11	0.10	0.07	0.06	0.07
Glass, large heavy plate..........................	0.18	0.06	0.04	0.03	0.02	0.02
Glass, ordinary window	0.35	0.25	0.18	0.12	0.07	0.04
Gypsum board, nailed to 2 by 4 studs on 16-inch centers	0.29	0.10	0.05	0.04	0.07	0.09
Plaster, gypsum, or lime, smooth finish on tile or brick	0.013	0.015	0.02	0.03	0.04	0.05
Plywood, ⅜-inch	0.28	0.22	0.17	0.09	0.10	0.11
Air, Sabins per 1000-cu. ft........................	—	—	—	—	2.3	7.2
Audience, seated in upholstered seats, per sq. ft. of floor area	0.44	0.54	0.60	0.62	0.58	0.50
Wooden pews occupied, per sq. ft. of floor area.....	0.57	0.61	0.75	0.86	0.91	0.86
Chairs, metal or wooden, seats unoccupied.........	0.15	0.19	0.22	0.39	0.38	0.30

Coefficients above were obtained by measurements
in the laboratories of the Acoustical Materials Association. Coefficients for other materials may be obtained from Bulletin XXII of the Association.

and carpeting. These materials are capable of having high absorption values at high frequencies, thus allowing the high-frequency energy in a room to be carefully controlled.

**Fig. 3-12. High-
frequency
absorption.**

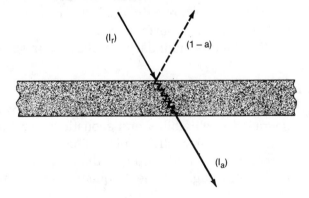

Low-Frequency Absorption

It can be seen from Table 3-2 that materials offering a degree of absorption in the high frequencies often offer little resistance to the low-frequency end of the spectrum, and vice versa. This is due to the fact that low frequencies are damped best by the compliance, or ability, to bend and flex with the incident waveform. This allows a portion of the waveform's power to be absorbed by the amount of energy required to flex the compliant material (Fig. 3-13).

Fig. 3-13. Low-frequency absorption.

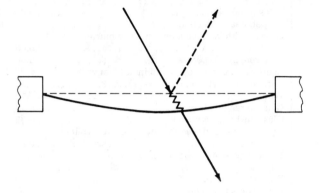

In order to determine the amount of absorption obtained by the total number of absorbers within a total volume area, at a specific frequency, it is necessary to calculate the average absorption coefficient (-a-) for all the surfaces together. This is found by:

$$-a- = \frac{s1\ a1\ +\ s2\ a2\ +\ \dots\ sn\ an}{S} \qquad \text{(Eq. 3-3)}$$

where,
  s1,2...n are the individual boundary surface areas in square feet,
  a1,2...n are the individual absorption coefficients of the individual boundary surface areas,
  S is the total boundary surface area in square feet.

Reverberation

A third criterion for a studio is that an allowance be made for a high degree of intelligibility and acoustic separation, both within the room and at the microphone pickup points. These are governed by the careful control and tuning of the reverberation constants, also known as T60, throughout the frequency spectrum of the studio environment.

Reverberation is the time required for a sound to die away to one millionth of its original intensity, resulting in a decrease of 60 dB, as shown by the formula:

$$T60 = \frac{V \times 0.049}{AS} \qquad \text{(Eq. 3-4)}$$

where,

    T is the reverberation time in seconds,
    V is the volume of the enclosure in cubic feet,
    A is the average absorption coefficient of the enclosure,
    S is the total surface area in square feet.

Reverberation is the persistence of a signal, in the form of reflected waves, after the original sound has ceased. This reflected persistence plays an extremely important role, both in the enhancement of our perception of music and in proper studio design.

As can be seen from Equation 3-4, reverberation time is directly proportional to two major factors: the volume of the room and the absorption coefficients of the studio surfaces. A large environment with a relatively low absorption coefficient, such as a concert hall, will yield a relatively long T60 decay time, while a small studio incorporating a heavy amount of absorption will give a very short T60 time.

The style of music and the room application will determine the optimum T60 for an acoustical environment. Fig. 3-14 shows a basic guide to reverb times for differing applications and musical styles. The typical recording studio may have a reverberation time ranging from 0.25 second, in a smaller absorbative studio environment, to 1.6 seconds in a larger music scoring studio. In certain designs, the T60 of a room may be controllable to fit the desired application by use of movable panels, louvers, or by the placement of carpets in a room (Fig. 3-15). Other designs may have sections of studio environment that will exhibit differing reverb constants, often known as the *live end* and *dead end* of a room. This permits the live end to be used in bringing certain instruments to life, such as strings, which rely heavily on room reflections and reverberation to bring them to life. At the same time, an acoustic environment is available which offers isolation within the studio for the louder percussive instruments, such as those found in the rhythm section.

Isolation between differing instruments and microphone channels is of extreme importance in the studio environment. If leakage is not controlled, the room's effectiveness for a multitude of applications becomes severely limited. In the studio designs of the 1960s and 1970s, this brought about the rise of the "sound sucker" era in studio design, which raised the room's absorption coefficient to an almost anechoic condition. With the advent of the new music of the 1980s, and a return of the respectability of live studio acoustics, the modern recording studio and the control room have seen an increase in

Fig. 3-14. Reverberation times for various types of auditoriums, at 512 Hz.

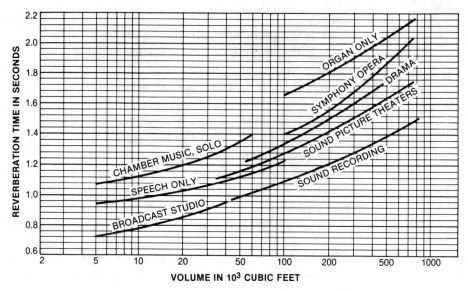

Fig. 3-15. Movable wall louvres or partitions, to control reflected sound in the studio.

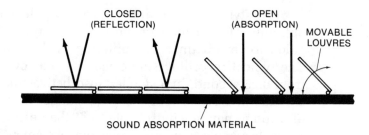

liveness and in physical dimensions, with a corresponding increase in the room's T60. This has added to the return to the thick, live-sounding, music production of earlier decades when studios were larger, more live, acoustic structures.

References

1. Don and Carolyn Davis, *Sound System Engineering*, Indianapolis: Howard W. Sams & Co., Inc., 1975.
2. Olson, Harry F., *Music, Physics, and Engineering*, New York: Dover Publications, Inc., 1967.
3. Tremaine, Howard M., *The Audio Cyclopedia*, Indianapolis: Howard W. Sams & Co., Inc., 1978.

4 MICROPHONES AND PICKUPS

A microphone is very often the first device found in the recording chain. In Chapter 1, we learned that the microphone (mike) is a transducer in that it changes one form of energy (sound waves) into another corresponding form of energy (electrical impulses). Given the fact that a microphone is a transducer, it can be a very subjective device, very much like a guitarist with eight guitars, each with a specific application. Each type and design of microphone has its own particular sound characteristics.

Three types of microphones used in recording today: dynamic, ribbon, and condenser.

The Dynamic Microphone

A *dynamic* or *moving coil* microphone is a pressure-operated device. It consists of a finely wrapped coil of wire which is attached to a delicate *diaphragm* and suspended in a permanent magnetic field (Fig. 4-1). When sound pressure waves hit the diaphragm, the diaphragm moves the coil in proportion to the wave intensity and causes the coil to cut across the fixed lines of magnetic flux supplied by the permanent magnet. Whenever metal wires (as in the coil) cuts across magnetic flux lines, a current flow is induced in the wire and an electrical output is generated. The amplitude of the current flow induced in the coil is proportional to the number of lines of flux cut by the coil (that is, how far it moves from its resting or no-signal position) and the speed at which the coil cuts the lines. The frequency of the signal is determined by how often the diaphragm reverses its direction of travel.

Basic pressure mikes are inherently omnidirectional. This means that they are equally sensitive to sound waves coming from any direction (Fig. 4-2). Dynamic microphones can also be designed to be directional. This means that they will be more sensitive to signals

Fig. 4-1. The components of a dynamic microphone.

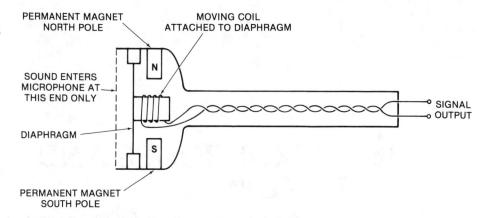

PERMANENT MAGNET
NORTH POLE

MOVING COIL
ATTACHED TO DIAPHRAGM

N

SOUND ENTERS
MICROPHONE AT
THIS END ONLY

SIGNAL
OUTPUT

DIAPHRAGM

S

PERMANENT MAGNET
SOUTH POLE

arriving from one direction than from another. The most common of these directional microphones is the *cardioid* mike which gets its name from its heart-shaped pickup pattern (Fig. 4-3). In the cardioid pickup pattern, signals arriving from the rear of the mike are acoustically phase shifted within the body of the microphone by a rear port and are applied to the back of the diaphragm so that signals arriving from the rear of the mike will be applied equally to both the front and back of the diaphragm at the same point in time, thus generating no output (Fig. 4-4A). In Fig. 4-4B, a signal from the front which enters the rear ports is phase shifted twice; once by the time it takes the wave to travel the external distance from the diaphragm to the port and a second time by the phase-shifting network inside the port. When the wave reaches the back of the diaphragm, it is in phase with the wave at the front and reinforces it. Between the front and rear source locations, the signal entering the port partially cancels and partially reinforces the signal at the front of the diaphragm, producing the cardioid pickup pattern.

Fig. 4-2. A pressure-operated mike is sensitive to sound waves originating from any direction.

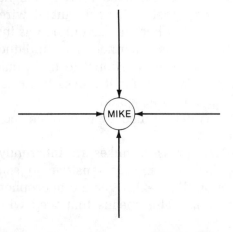

MIKE

Fig. 4-3. A cardioid pickup pattern.

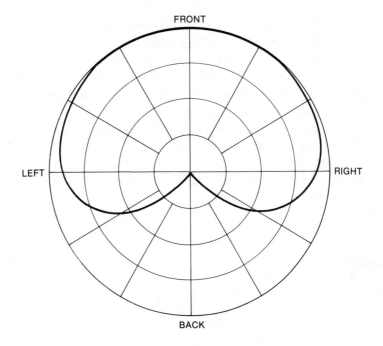

Fig. 4-4. Directional operation of the cardioid microphone.

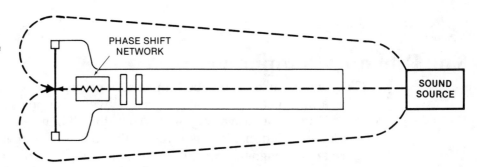

(A) Signals arrive at the front and rear of the diaphragm simultaneously.

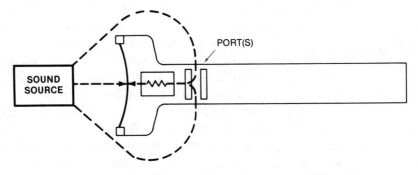

(B) Signal through port is shifted to be in phase with front signal.

The attenuation of an equal signal, which is set at the same distance from a cardioid mike but 180° off-axis with respect to an on-axis front signal, is called the *front-to-back discrimination*. The axis of a mike is the imaginary line drawn perpendicular to the plane of the diaphragm (Fig. 4-5). Other directional designs, called *hypercardioid* and *supercardioid*, sacrifice some front-to-back discrimination in favor of a narrower pickup pattern (Fig. 4-6). Fig. 4-6B shows the frequency response on-axis, 180° off-axis, and 150° off-axis. Regular cardioids have maximum off-axis rejection at 180°, while supercardioids, such as the Electro-Voice RE-16 (Fig. 4-7), have maximum rejection at 150°.

Fig. 4-5. A microphone axis is the line perpendicular to the diaphragm on the side exposed to the sound waves.

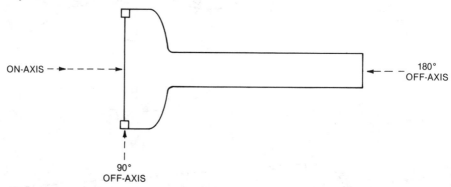

The Ribbon Microphone

The *ribbon* mike, also called a *pressure gradient* or *velocity* mike, utilizes a thin metal ribbon suspended between the poles of a magnet to sense the sound wave (Fig. 4-8). When the ribbon is moved by sound pressure variations, it cuts through the lines of flux generated by the permanent magnet, and this induces a voltage into the ribbon. This voltage becomes the signal output. The mike is called *pressure gradient* because the motion of the ribbon is determined by the difference in pressure between its front and back sides. The difference is proportional to the velocity of the air molecules which make up the wave, and thus, its third name, *velocity* mike.

Since the metal ribbon is exposed to sound waves at both its front and rear sides (Fig. 4-9A), it is equally sensitive to sounds coming from both directions with sound from the rear producing a voltage that is 180° out of phase with the voltage produced by the sound from the front. Sound waves 90° off-axis will produce equal, but opposite, pressures at the front and rear of the ribbon (Fig. 4-9B) and will cancel. As a result, the ribbon mike is inherently bidirectional, with a figure-eight-shaped pickup pattern (Fig. 4-10).

Fig. 4-6. The Electro-Voice RE-16 supercardioid pickup pattern (*Courtesy Electro-Voice Inc.*).

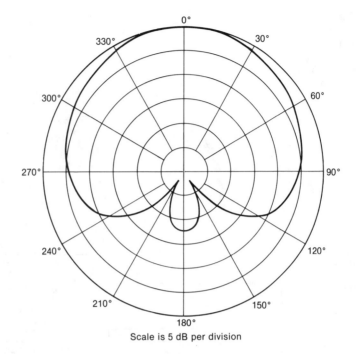

Scale is 5 dB per division

(A) Pickup pattern.

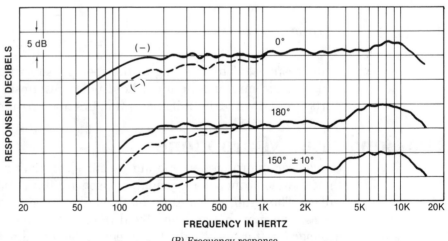

(B) Frequency response.

Other directional patterns may be obtained with a ribbon microphone by closing off the acoustic path to the rear of the ribbon by varying degrees (Fig. 4-11). This is especially true of older ribbon mikes. If the acoustic paths to the rear of the mike are completely closed off (Fig. 4-11A), the ribbon mike will exhibit an omnidirec-

Fig. 4-7. The
Electro-Voice
RE-16 dynamic
supercardioid
microphone
(Courtesy Electro-
Voice Inc.).

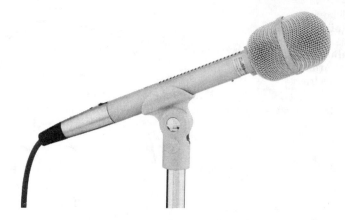

Fig. 4-8. The
basic
components of a
ribbon
microphone.

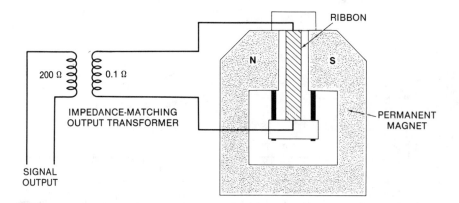

tional response. The opening of the rear phase shift path (Fig. 4-11B)
will give the microphone a cardioid polar response, while with both
ports open, it will exhibit its natural figure-eight pattern (Fig. 4-11C).

The Condenser Microphone

Condenser microphones operate on an electrostatic principle rather
than the electromagnetic principle used by the dynamic and ribbon
mikes. The *head*, or *capsule*, of the mike consists of two very thin
plates, one movable and one fixed (Fig. 4-12). These two plates form
a *capacitor* (formerly called *condenser*; hence, the name condenser
microphone). A capacitor is an electrical device which is capable of
storing an electrical charge. The amount of charge a capacitor is able
to store is determined by its value of capacitance and the applied
voltage, according to the formula:

$$Q = C \, V$$ **(Eq. 4-1)**

Fig. 4-9. Sound sources on-axis and 90° off-axis at a ribbon mike.

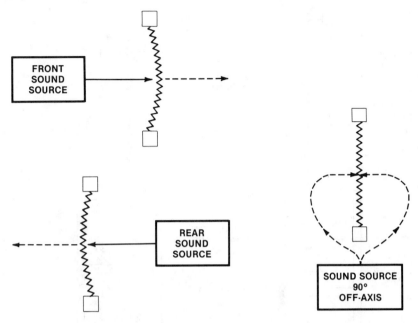

(A) The ribbon is sensitive to sounds at the front and rear.

(B) Sound wave from 90° off-axis.

Fig. 4-10. The normal pickup pattern of a ribbon microphone.

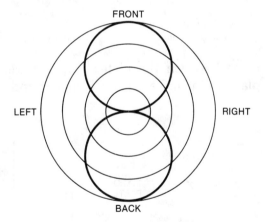

where,

       Q is the charge, in coulombs,
       C is the capacitance, in farads,
       V is the voltage, in volts.

    The capacitance of the capsule is determined by the composition and surface area of the plates (which are fixed values), the *dielectric* or substance between the plates (which is air and fixed), and the dis-

Fig. 4-11. A method of producing variable pickup patterns from a ribbon microphone using ports and an acoustical phase-shift network.

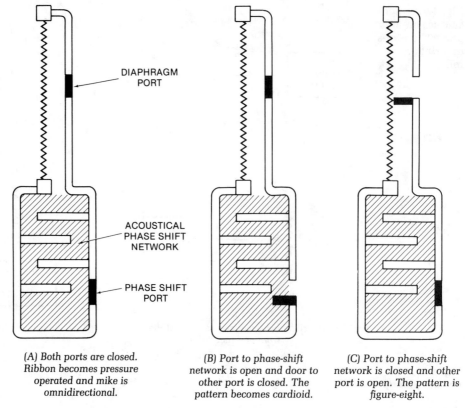

DIAPHRAGM PORT

ACOUSTICAL PHASE SHIFT NETWORK

PHASE SHIFT PORT

(A) Both ports are closed. Ribbon becomes pressure operated and mike is omnidirectional.

(B) Port to phase-shift network is open and door to other port is closed. The pattern becomes cardioid.

(C) Port to phase-shift network is closed and other port is open. The pattern is figure-eight.

tance between the plates (which varies with sound pressure). The plates of the mike capsule form a sound-pressure-sensitive capacitor.

Fig. 4-12. In a condenser microphone capsule, the sound pressure varies distance d between the plates.

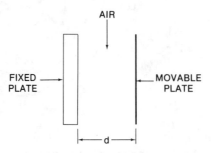

AIR

FIXED PLATE

MOVABLE PLATE

d

In the design used by the majority of manufacturers, the plates are connected to opposite sides of a dc power supply which provides a *polarizing voltage* for the capacitor (Fig. 4-13). Electrons are drawn from the plate connected to the positive side of the power supply and forced through a high-value resistor onto the plate connected to the negative side of the supply. This continues to occur until the charge on the capsule (that is, the difference between the number of electrons on the positive and negative plates) is equal to the capacitance of the capsule times the polarizing voltage. When this equilibrium is reached, no further appreciable current flows through the resistor. If the mike is fed a sound-pressure wave, the capacitance of the head changes. When the distance between the plates decreases, the capacitance increases, and when the distance increases, the capacitance decreases.

Fig. 4-13. The power supply charges the capacitor through a high-value resistor.

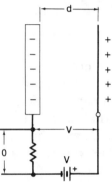

According to Equation 4-1, Q, C, and V are interrelated, so if charge (Q) is constant and sound pressure moves the diaphragm changing capacitance (C), the voltage (V) must change in direct proportion in order for the equation to hold true.

The high-value resistor, in conjunction with the capacitance of the plates, produce a circuit *time constant* that is longer than a cycle of an audio frequency. The time constant of a circuit is a measure of the time needed for a capacitor to charge or discharge. Since the resistor prevents the capacitor charge from varying with the rapid changes in capacitance caused by the applied sound pressure, the voltage across the capacitor must change according to $\Delta V = Q/\Delta C$ (Fig. 4-14). The resistor and capacitor are in series with the power supply, so the sum of the *voltage drops* across them must equal the supply voltage. When the voltage across the capacitor changes, the voltage across the resistor changes equally, but in the opposite direction. The voltage across the resistor then becomes the output signal. Since the signal off of the diaphragm is a low-level and extremely high-impedance signal, it is amplified within the body of the microphone to prevent the hum, noise pickup, and signal-level losses that would occur (due to the resistance of cables and other factors) if the amplifier was at a distance from the capsule. This mike preamplifier is another reason that many condenser microphones need power supplies.

Fig. 4-14. If a sound wave decreases the plate spacing by Δ d, the capacitance increases by Δ C and the voltage across the plates falls by Δ V.

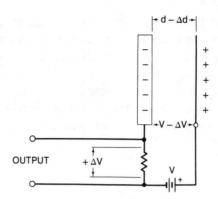

Electret microphones (Fig. 4-15) operate upon the same principle of capacitance except that the polarizing voltage is permanently stored in the capsule plates in the form of an electrostatic charge so that no external power supply is necessary. The high impedance output still necessitates the use of an amplifier to boost level and lower impedance, however, so an internal battery power supply is often required.

Fig. 4-15. Diagram of an electret condenser microphone. Its circuit is equivalent to that of Fig. 4-13.

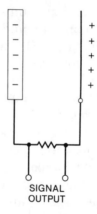

In recent years, a new form of electret condenser mike, the *Pressure Zone Microphone* or *PZM*, has been gaining popularity and recognition (Fig. 4-16). The PZM is constructed of a flat surface plate on which is mounted an encased diaphragm. This diaphragm is open to the plate by way of a thin slitted port. The sound pressure waves are accumulated across the horizontal plane of the plate and collected at the diaphragm. The diaphragm is mounted in the "pressure zone" just above the plate, a region where the direct and reflected waves effectively add in-phase, adding to the mike's *microphone sensitivity* by 6 dB.

Fig. 4-16. The Pressure Zone Microphone *(Courtesy Crown International).*

The theory behind the PZM rests with the fact that, with conventional microphone techniques, sound travels to the mike via two paths: directly from the source to the microphone and as a reflection off a nearby surface (which is generally delayed with respect to the direct sound). When this direct and delayed sound combines at the microphone, the result is a phase cancellation at various frequencies. Since the PZM microphone can be placed directly on the surface, which otherwise would provide the reflective path, the combining effect of the phase cancellation is eliminated, resulting in a smoother frequency response pickup.

A condenser design different from the electret mike is used by Sennheiser Electronic Corp. in their microphones. Rather than applying a polarizing voltage to the capsule and sensing its varying charge-holding capability, the capacitor is used as part of the tuning circuit for a high-frequency oscillator. Variations in sound pressure cause frequency modulation of the oscillator (variations in oscillator frequency proportional to the intensity of the sound waves), resulting in a signal similar to that broadcast by an fm radio station. The high-frequency signal is converted into audio the same way that an fm radio signal is. The lack of a polarizing voltage makes the mike less sensitive to physical shock and changes in humidity or temperature which can cause arcing of the polarizing voltage between the plates.

Condenser mikes are inherently pressure operated and, thus, omnidirectional. Other directional patterns can be achieved through the use of a perforated fixed plate and a second movable plate in the capsule. By placing the two movable plates on opposite sides of the fixed plate, two capacitors are formed. By polarizing the two movable plates to one polarity and polarizing the fixed plate to the opposite polarity (Fig. 4-17), an omnidirectional pickup pattern is achieved. As the intensity of the voltage on one of the movable plates is decreased while holding the polarization intensity of the other plate unchanged (Fig. 4-18), the pickup pattern becomes more and more cardioid. When the movable plates are at opposite polarities and the fixed plate is set at ground potential (Fig. 4-19), a figure-eight pattern is achieved. Selectable pattern switches are generally provided on the outer body of many condenser mikes.

Fig. 4-17. An omnidirectional condenser microphone diagram.

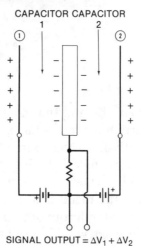

Fig. 4-18. A cardioid condenser microphone.

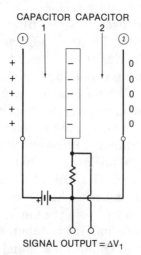

Fig. 4-19. A figure-eight condenser microphone. Operation is the same as in Fig. 4-17 except that capacitors 1 and 2 produce voltages across the resistor that are 180° out of phase with each other.

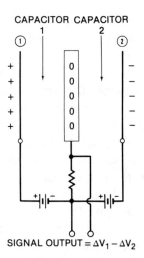

CAPACITOR CAPACITOR

SIGNAL OUTPUT $= \Delta V_1 - \Delta V_2$

Because condenser mikes have built-in amplifiers, many of them also have built-in *attenuation pads* immediately after the capsule output. This prevents the preamp contained within the mike from overloading when exposed to very high sound-pressure levels. If the capsule output level were high enough to overload this amp, the output signal fed to the mike preamp in the control room would be distorted, and no amount of control room preamp padding would solve the problem. The pad within the mike itself avoids this problem and should be used if the signal to be picked up is very loud.

Directional mikes have the property of their bass response increasing as the signal source gets closer to the mike. This *proximity effect* is noticeable when the source is closer than about 2 feet from the mike, and the effect increases as the source gets nearer. The effect is caused by an increase in pressure gradients at short mike-to-source distances. At 2 feet, the sound intensity is four times the intensity present at 4 feet from the source, while, at 1 foot, the intensity of the sound is sixteen times that present at 4 feet. This increase in pressure gradient occurs throughout the audio spectrum, but there is a greater phase change between the front and the rear of the mike at the higher frequencies due to their shorter wavelengths. This causes some of the pressure gradient to be cancelled for higher frequencies, producing a bass boost. This boost is somewhat greater in bidirectional mikes than in cardioid mikes. To compensate for this effect, a bass rolloff switch is often provided to reduce the bass response back to flat. In microphones designed for close-up work, the frequency response may be rolled off at the low end, and the proximity effect is used to restore the bass response. Another way of reducing the proximity effect and the associated popping of the letter "p" is to choose an omnidirectional microphone for the existing application.

Each of the three styles of microphones has its own individual characteristics. Dynamic mikes tend to be more rugged than their ribbon or condenser counterpart, although they too, in recent years, have become quite rugged. Older ribbon mikes may be easily damaged by physical shock or wind shock which can damage, deform, or break the ribbon. The dynamic and ribbon types do not need power supplies to operate and are not as sensitive to changes in humidity or temperature as condenser mikes. Condenser mikes, on the other hand, have better transient response (faster response to rapid changes in sound pressure), flatter frequency response (especially in the high end), and much higher output levels due to their built-in amplifiers. Although ribbon mikes are still in use today, they were more popular during the early days of radio. Only a few microphones are now being produced using ribbons.

Specifications

In choosing a microphone, there are specific data which can assist you in making the right choice for a particular application. While some technical information may be supplied in the manufacturer's specification sheets, vital information will often not be included. The best gauge of a microphone's performance will, in the long run, be its sound.

The *frequency response* of a microphone can be valuable information. It will give you clues as to how the mike will react at specific frequencies. Some mikes may exaggerate the highs and others the middle frequencies, while still others may have a flat response curve. Each of these may be the best choice for different applications.

A significant piece of data which presently has no accepted standard of measure is the *transient response* of a microphone (Fig. 4-20). Transient response is the measure of how quickly the diaphragm of a mike will react to a waveform. This figure varies wildly among microphones and is a major reason for the difference in sound quality between the three types. The diaphragm of a dynamic microphone can be quite large (up to $2\frac{1}{2}$ inches), and with the addition of the coil of wire and its core, the combination can make for a very large mass when compared to the power of the sound wave which drives it.

**Fig. 4-20.
Characteristic
transient
response of the
three
microphone
types.**

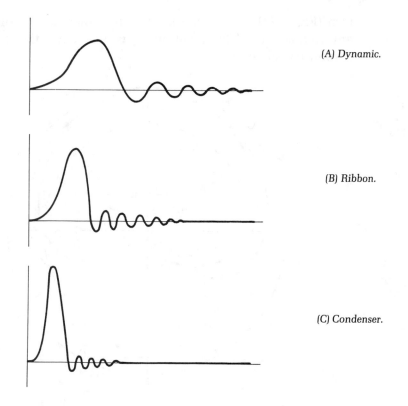

(A) Dynamic.

(B) Ribbon.

(C) Condenser.

A dynamic mike can be very slow in reacting to a waveform, giving a rugged, gutsy sound. The diaphragm of a ribbon mike is much lighter by comparison, being made of a thin, corrugated ribbon. This means that its diaphragm can react more quickly to a sound waveform, resulting in a clearer sound. Older ribbon models tend to have large grills covering the diaphragm to protect it against wind. These grills often tend to cancel out the higher frequencies, giving these ribbon mikes a mellow quality. The condenser mike, on the other hand, has an extremely light diaphragm which varies in diameter from $2\frac{1}{2}$ inches down to $\frac{1}{4}$ inch and with a thickness of about 0.0015 inch. This means that the diaphragm offers little mechanical resistance to a sound pressure wave, allowing it to track the wave more accurately over the entire range.

A significant piece of data is the *directional characteristic* of the mike. However, this can be meaningless without a curve of frequency versus number of degrees off-axis. All directional mikes have a different frequency response off-axis than they do on-axis, but better mikes maintain the on-axis characteristics over a wider pickup angle. The directional information is given on a circular polar pattern (Fig. 4-21), with differently coded lines corresponding to different frequencies. The ideal mike would have the same frequency response at all angles.

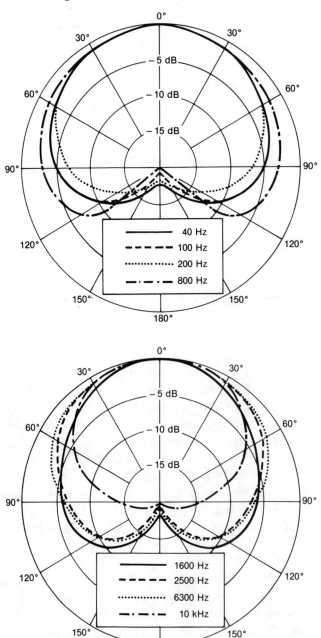

Fig. 4-21. Polar response curves for the SM53 microphone *(Courtesy Shure Brothers, Inc.).*

Next are the mike *sensitivity* and *self-noise level*. These figures determine the best possible microphone signal-to-noise ratio. The higher the output level, the less amplification is needed from the preamp. Using a variable-gain preamp, gain can be set to exactly that needed so as to not unnecessarily amplify thermal noise. Another factor that must be considered is if a high-output mike is used without a variable-gain preamp, mike pads must be available in order to prevent the preamp from being overloaded on loud passages. The sensitivity is determined by measuring the mike output into its rated impedance load with a certain sound-pressure level at the diaphragm of the mike. In comparing the sensitivity of two mikes, they must both receive the same sound-pressure level at the same frequency. The most common sound-pressure level used is 10 dynes/cm^2 (10 microbars) at 1 kHz and the rating is given in dB re 1 milliwatt (dBm). The self-noise level is the noise level generated by the mike when no sound reaches it. In a dynamic or ribbon mike, this is the thermal noise caused by electrons moving in the coil or ribbon, while, in condenser mikes, it is mainly due to the noise in the built-in preamp. For any given mike, this noise cannot be eliminated, and it places a restraint on how good the signal-to-noise ratio can be.

Probably the least important specification for recording use is the level at which the mike begins to create harmonic distortion. This is relatively unimportant because all modern mikes can withstand very intense sound pressure without creating appreciable harmonic distortion. The rating is usually given in dB of sound-pressure level for 0.5% total harmonic distortion (thd).

Microphone Impedance

Microphones are available with different output impedances. *Output impedance* is a rating used to match the signal-providing capability of one device with the signal-drawing (input impedance) requirements of another device. Impedance is measured in ohms and its symbol is Z. Commonly used microphone output impedances are 50 ohms, 150 to 250 ohms (low), and 20 to 50 Kohms (high). Each impedance range has its advantage. In the past, high-impedance mikes were less expensive to use because the input impedance of tube-type amplifiers was high. To be used with low-impedance mikes, they required expensive input transformers. All dynamic mikes, however, are low-impedance devices, and those with high-impedance outputs achieve them through the use of a built-in impedance step-up transformer. A disadvantage of high-impedance mikes is the susceptibility of their high-impedance mike lines to the pickup of electrostatic noise, such as that caused by fluorescent lights and motors. This makes the use of shielded cable necessary. In addition, the use of a conductor surrounded by a shield creates a capacitor which is, in effect, connected across the output of the microphone. As the length of the cable increases, the capacitance increases until, at about 20 to 25 feet, the cable capacitance begins to short out much of the high-frequency information picked up by the mike. Thus, good results with high-impedance mikes are limited to use with cable lengths of 25 feet or less.

Very-low-impedance (50 ohms) mikes have the advantage that their mike lines are fairly insensitive to electrostatic pickup. They are sensitive, however, to induced hum pickup from electromagnetic fields, such as those generated by ac power lines. This pickup can be eliminated by using twisted-pair cable. The currents magnetically induced in this cable will flow in opposite directions and will cancel each other out in the input transformer of the mike preamp. The 50-ohm lines do not need shielding, for this is effective only against electrostatic, not electromagnetic pickup. Cable lengths are limited to about 100 feet due to signal power losses resulting from cable resistance. These losses, however, do not degrade frequency response, merely signal-to-noise ratio. They can be reduced by using larger conductors which have less resistance.

The 150- to 250-ohm mike lines are less susceptible to electromagnetic pickup than 50-ohm lines but are more susceptible than high-impedance lines. They are also more susceptible to electrostatic pickup than 50-ohm lines, but are less susceptible to this type of pickup than high-impedance lines. As a result, shielded twisted-pair cable is used and lowest noise is attained through the use of *balanced* input circuits (Fig. 4-22). This means that two wires carry the signal voltage and a shield which is connected to ground is wrapped around them. Neither of the two signal lines are grounded. High-impedance mike lines use *unbalanced* circuits in which one signal lead goes through the center of the cable and is surrounded by a shield which is grounded and used as the second signal lead. Balanced lines operate on the principle that audio works with alternating current between the two leads and an electrostatic or electromagnetic pickup will be presented to both leads simultaneously. These equal charges will, upon arrival at the input transformer or balancing amplifier, be of equal and opposite levels, resulting in the cancellation of the unwanted signal while passing the audio signal unaffected. The 150- to 250-ohm mike lines have low signal losses and can be used with cable lengths up to several thousand feet. The mike lines used in most recording studios are 200-ohm balanced lines with the shield grounded at the preamp end only. This is done to prevent *ground loops* which can produce hum in the signal if shields are grounded at more than one point.

**Fig. 4-22.
Diagram of
typical
microphone/
XLR-cable
connections.**

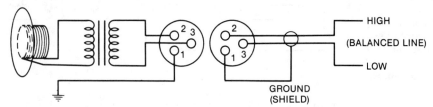

Phantom Power
============

Most present-day condenser microphones can be powered directly from the console by way of a *phantom* power supply which provides power to all console mike inputs simultaneously. This phantom powering system does not interfere with the operation of dynamic mikes. The positive side of the power is fed to both sides of each mike line through identical value resistors so that there is no voltage differential between the two signal-carrying leads (Fig. 4-23). The negative side of the supply is connected to the shield. Condenser mikes designed to make use of this voltage no longer need internal batteries, external battery packs, or individual ac-operated power supplies. The resistors used in connecting the phantom supply to the mike inputs provide isolation between the two conductors of each mike line and between mikes. Shorting either of the signal leads to ground on one input does not affect the phantom power delivered to the other inputs. However, if two or more inputs are shorted, the phantom voltage may drop too low to be usable. This can happen if mike cables become defective and one or both signal leads short to the shield. The size of the resistors used in the phantom supply depends upon the voltage and current requirements of the microphones to be powered.

Direct Injection
============

When an instrument being recorded has an electrical output position, such as that in an electric guitar or synthesizer, it is possible to record that instrument direct without the use of a microphone. If the instrument is being played in the studio, a *direct injection* or *D.I.* box (Fig. 4-24) would most likely be used. The output of the instrument is

Fig. 4-23. The phantom powering system permits simultaneous operation of phantom-powered condenser microphones and dynamic or ribbon microphones.

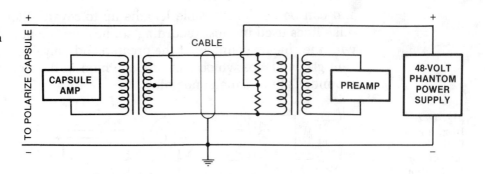

plugged into the input of the D.I. box and the microphone-level XLR output is plugged into the console and assigned to a track as if it were a microphone input.

Fig. 4-24. The "DI-100" is a 17-ounce (480 gram) precision Direct Box, with adjustable gain, for high-end professional use *(Courtesy Artists X-ponent Engineering).*

The D.I. box serves two purposes; it reduces the line level of an instrument down to the microphone level and, if necessary, it electrically isolates the instrument from the console by way of an internal transformer or amplifier. This will electrically isolate the console from the instrument and, possibly, eliminate *buzz* which can result from differing ground potentials.

The musician may monitor his instrument via headphones directly from the console's *cue send*, eliminating any need for an instrument amplifier, or he can take the instrument feed from the D.I. box and plug it into his amplifier in the studio. The latter option gives the choice of setting up a mike on the amplifier and mixing the direct and "miked" signals together in the mix.

There are two basic designs of D.I. boxes: transformer and active. The first design isolates and steps down the signal by way of a transformer. Transformers often are associated with a degraded transient response and frequency ringing. The active D.I. box does not add any appreciable sound of its own and, since it contains an amplifier, it must be powered by battery or phantom powering.

Pickups

Most pickups used for recording are contact pickups. These devices are most often attached to the instrument bridge or sounding board of instruments such as guitars, violins, pianos, and more recently, drums. Contact pickups are often invaluable for live public address application.

The most common type of contact pickup is one which uses a crystal as a transducer. A thin crystal slice, which has a natural polarized electrical potential, will generate an electrical impulse when it is bent. The contact pickup is designed so that the vibrating surface of the instrument will correspondingly stress the crystal, thus producing an electrical output.

References

1. Alexandrovich, George, "The Audio Engineer's Handbook," *db, The Sound Engineering Magazine*, Vol. 4, No. 8, August 1970, pp. 4-6.
2. Bartlett, Bruce, "PZM—A Useful Recording Tool," *Tape Deck Magazine*, 1984.
3. Eargle, John, "How Capacitor Mics Produce Cardioid Patterns," *db, The Sound Engineering Magazine*, Vol. 5, No. 4, April 1971, pp. 32-34.
4. Malo, Ron, "Phase and the Single Microphone," *Recording engineer/producer*, Vol. 2, No. 6, November/December 1971, p. 17.
5. McCulloch, John A., "The Feedback Loop," *db, The Sound Engineering Magazine*, Vol. 2, No. 1, January 1968, pp. 2-4.
6. Nisbett, Alec, *The Technique of the Sound Studio*, New York: Hastings House Publishers, Inc., 1972.

7. Tremaine, Howard M., *The Audio Cyclopedia*, Indianapolis: Howard W. Sams & Co., Inc., 1978.
8. Woram, John M., "The Sync Track," *db, The Sound Engineering Magazine*, Vol. 5, No. 2, February 1971, pp. 12-14.

5 MICROPHONE TECHNIQUES

We have learned that differences in microphone types, and even minor differences in microphone design, can give a mike a separate and distinctive sound character. There are, therefore, many different mikes for different applications and it is up to the engineer to choose the right one for the job—but this is really only half the story. Not only is the right choice of microphone important but the right placement of the mike can be just as important to getting the right sound.

It should be emphasized from the start that microphone placement is an art form—an engineer's tool. What is considered as a bad technique now may be acceptable as the standard placement technique five years from now. As new music styles develop, new studio recording techniques also tend to evolve. This helps give the music a new sound and it ties in with an old adage that has been taught for years and which is called the first rule of recording:

First Rule of Recording: *There are no rules.*

Recording is an artform and as such is totally open to change and experimentation; this is what keeps both the music and the industry alive and fresh. There are no rules in microphone techniques, only guidelines. It is the goal of this chapter to present a few of these guidelines.

Distant and Close Miking

In modern recording, there are basically two styles of microphone techniques: distant miking and close miking (Fig. 5-1). *Distant miking* refers to the placement of one or more microphones at a distance of generally five or more feet. This style of recording serves a dual function: to place the mike at such a distance that the entire instrument sound is picked up, thus preserving the overall tonal balance of the instrument. Another reason may be to place the microphone such that the instrument's acoustic environment is included in the mike's pickup area and is combined with the direct signal of the instrument. Distant miking is often used in the pickup of a large ensemble, such as a symphony orchestra or choral group, in a concert hall environment. The mike is placed at such a distance as to pick up the entire ensemble as well as the acoustics of the concert hall (Fig. 5-2). This placement treats the ensemble as one balanced instrument, leaving control in the hands of the conductor. The use of distant miking is not, by any means, limited to classical music as is sometimes believed but may be used in many different circumstances, ranging from the overall pickup of a string section in a large studio to the larger-than-life driving sounds of a drum being miked from a distance. Distant miking when used as an effect can add depth to a sound. For example, if a drum machine is played loudly over the studio monitors, and then recorded by a distant pair of mikes, the effect will be a fatter, deeper sound (Fig. 5-3).

Fig. 5-1. Distant (a) and close (b) microphone placement to a typical source.

Fig. 5-2. An ensemble in a room with a single distantly placed mike.

Fig. 5-3. Using studio monitors and a distant mike to get a deeper sound.

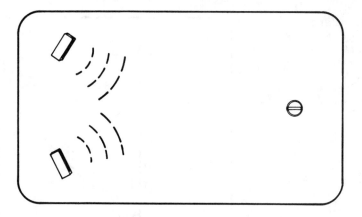

Close miking, conversely, refers to the miking of an instrument at the close proximity of 1 inch to 3 feet. Close miking is the method most often used in the modern multitrack studio. Miking an instrument at such a close range will serve to effectively exclude the acoustic environment from being recorded onto tape. Because sound diminishes with the square of its distance, a sound 6 feet from a mike will be insignificant compared to a sound originating 3 inches in front of it; thus, only the desired sound will be recorded onto tape. The acoustical environment is for all practical purposes not picked up by the mike. This goes for both the unwanted acoustic environment and unwanted nearby instruments. The latter factor is important in multitrack recording. Since the basic idea of modern multitrack recording is the control of individual tracks, isolation is important. If the sounds of one instrument (Fig. 5-4) were picked up by a nearby mike which is also recording another instrument, a condition known as leakage would be set up. Since the mikes contain both the direct sound and the leaked sounds from the other instrument(s), control, in mixdown, over one of these instruments would be difficult without affecting the level and sound character of the other. Needless to say, unwanted leakage is a condition avoided in the multitrack studio.

There are two ways of correcting the leakage problem illustrated in Fig. 5-4. Bringing the two microphones in closer to their respective instruments (Fig. 5-5A), and/or placing an acoustic barrier (known as a flat) between the two (Fig. 5-5B), will reduce the acoustic leakage.

Since close miking involves distances that commonly range from 1 to 6 inches, the total sound balance may not be picked up. Rather, the mike may be placed so closely to the musical instrument that only a small portion of the instrument's sound may be picked up, thereby

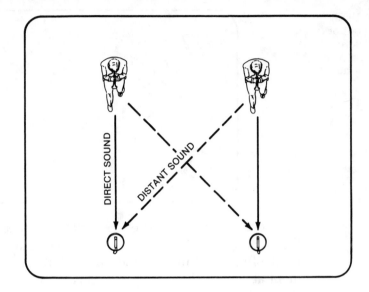

receiving only a selected balance. At such close distances, the movement of a mike only a few inches may change the tonal balance of the originating instrument's sound.

In review, *distant miking* techniques place the microphone at a distance of five or more feet from the sound source to preserve the overall tonal balance of the instrument, while also picking up the natural acoustics of the environment. *Close miking* uses working distances of usually less than three feet. This eliminates leakage caused by nearby instruments and the acoustical environment. As per the "first rule" of recording, the choice of style is up to you. A jazz flute can be made interesting, whether it was recorded in the studio at a distance of one foot (with added reverb) or in the Taj Mahal at a distance of fifty feet. In certain cases, the mix of these mediums may make for an interesting effect.

A word of caution must be given here. In the recording of a symphony orchestra, where distant miking techniques are fairly well established, a mix of distant and close miking techniques can be disastrous. If an orchestral piece were to call for a solo oboe passage, it is possible that the oboe would sound dull over the main microphones (point a in Fig. 5-6). This would call for an *accent* or *fill* mike that would be placed closer in towards the oboe. However, if the mike were placed just 1 foot (point b) from the instrument, the sudden jump from the distant mike to such a close distance would be too drastic a balance change. Whereas, placing the accent mike at a dis-

Fig. 5-5.
Reducing
leakage.

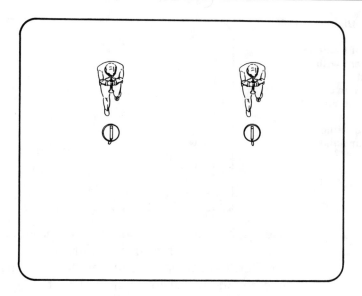

(A) Placing mikes closer to the sources.

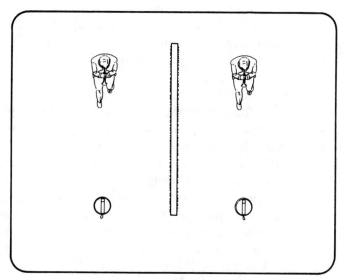

(B) Using an acoustical barrier (flat).

tance of about 12 feet (point c in Fig. 5-6) would balance out the solo instrument so that it would not be too noticeably out of balance when mixed in with the main stereo pair.

Phasing

Stereo effects can be produced by using two microphones to pick up the same sound source, as, for example, in close miking, when each

Fig. 5-6. Distant and close miking of a solo instrument with a distant microphone at point a and an accent mike located at either point b or point c.

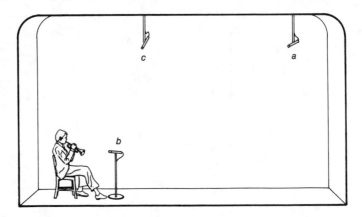

mike might pick up various timbres produced at different places on the instrument, such as over the bridge and over the hole of an acoustic guitar. Care must be taken in the use of two or more mikes on the same sound source to be certain that they are electrically in phase with each other. If they are not, frequency cancellations will occur which reduce the volume and change the sound character. In the case of out-of-phase mikes placed left and right in a mix, a mono listener (over a car radio for example) will hear very little of the instrument, while a stereo listener will hear the desired balance. A simple check for phase is to assign the output from both (or all) of the mikes to one speaker and listen for drastic changes in the sound.

If one or more mikes is 180° out of phase with the others, the problem may be in the miswiring of a mike cable (assuming the studio equipment was properly checked for phase integrity when wired up) and can be solved by either replacing the cable with a good one, by changing the polarity of the audio leads (pins 2 and 3) on the XLR connecter, by using a phase reversal adapter, or by reversing the phase on the console's input strip, when available.

Acoustical phase problems can be caused by using too many mikes near each other, where each picks up some of the sound picked up by the others. The different path lengths from source to mikes result in different phase relationships between the microphone outputs for the same sound. If these are combined at any point, cancellations occur producing severe dips in the frequency response of the signal. A rule of thumb to avoid acoustical phase cancellation when mikes are mixed is to make sure that the distance between any two mikes is at least three times the distance from each mike to its intended source.

If *distant miking* of an instrument is used to pick up some of the *room sound*, the placement of the distant mike at a random height often results in a hollow sound due to *phase cancellation* of the direct sound with the sound reflected from the floor (Fig. 5-7). The sound reflected from the floor travels farther than that which reaches the mike directly. Frequencies for which this extra path length is one half of a wavelength, or an odd integer multiple of one half a wavelength, arrive 180° out of phase with the direct sound, producing dips in the frequency response of the signal at the mike output. Since the reflected sound is at a lower level than the direct sound due to its travelling further and losing energy when it hits the floor, the cancellation is not complete.

Fig. 5-7. A distant mike picks up both the direct sound from the source and the reflected sound from nearby surfaces, such as the floor.

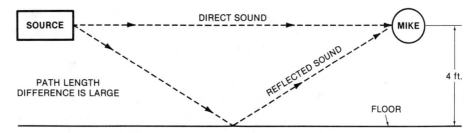

Moving the mike closer to the floor reduces the path-length difference (Fig. 5-8) and raises the frequency of cancellation. In practice, a mike height of ⅛ to ¹⁄₁₆ of an inch keeps the lowest cancellation frequency above 10 kHz. The design of the Pressure Zone Microphone (PZM) mentioned earlier places the diaphragm well within these height restrictions making it an ideal mike for this application.

Fig. 5-8. Moving the distant mike toward the reflecting surface decreases the difference in path length and raises the lowest frequency of cancellation.

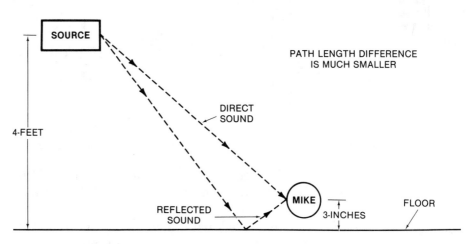

The placement of a conventional mike close to a reflecting surface increases its output by 6 dB. Experiments have shown that cancellations are a problem whenever mike-to-source distance is greater than one or two times the distance of the source to the reflecting surface. Surfaces such as the carpeted floor or the absorbent walls of a recording studio reduce the intensity of the reflected sound, thereby reducing the effect of the cancellations.

Stereo Miking Techniques

In speaking of stereo miking techniques, we shall refer to the use of two microphones in obtaining a stereo image. These techniques may be used in either close or distant overall miking of a large or small ensemble, of a single instrument on location or in the studio (such as a piano or guitar), or even of background vocals; the only limitation is your imagination. There are basically three methods of stereo miking using two microphones: the 3:1 principle, the XY technique, and the M-S method.

The *3:1 principle* states that in order to maintain phase integrity for every unit of distance between the mike and its source, the distance between the two mikes should be at least three times that distance (Fig. 5-9). Thus, if a pair of microphones were placed over the sound board of a piano at a distance of 1 foot, the separation between the two mikes should be at least 3 feet.

Fig. 5-9. Using the 3:1 principle to position microphones over the sound board of a piano.

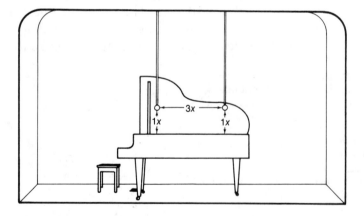

The *XY stereo miking technique* is an intensity-dependent system as it uses only the *cue* of amplitude to discriminate direction. With the XY technique, two microphones of the exact same type and manufacture are placed with their grills as close together as possible (without touching) and facing at right angles to each other. The midpoint, or 45° angle, between the two mikes is then faced toward the musical instrument and the mike outputs are panned left and right

(Fig. 5-10). Even though the two mikes are placed together, the stereo imaging is excellent—often better than when using the 3:1 method. There is also the added advantage of having no appreciable phase problems due to the close proximity of the mikes. The angles of the two mikes may be changed to fit the occasion, with the preferred angles being from 90° to 120°, with the generally accepted polar pattern being cardioid, although two crossed figure-eight mikes are capable of producing excellent ambient results.

Fig. 5-10. XY stereo miking technique pattern.

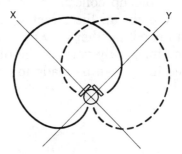

Currently, there are stereo microphones on the market that contain two diaphragms in the same microphone case, with the top diaphragm generally being rotatable by 180°, which accommodates them for XY techniques (Fig. 5-11).

Fig. 5-11. AKG C422 stereo mike *(Courtesy North American Phillips Corp.).*

The *M-S* or *mid-side* method (Fig. 5-12) is similar to the XY technique in that it works off of two diaphragms in close proximity and most often with a stereo mike like the one shown in Fig. 5-11. The mid-side method differs from the previous two methods in that it requires an external transformer matrix to operate in conjunction with the microphones. In the classic M-S system, the microphone that is designated as the *mid microphone* has a cardioid pattern which is oriented towards the sound source. The other microphone has a figure-eight pattern facing sideways, with the null side of the pattern coinciding with the main axis of the cardioid pattern. This *side (S)* pickup collects the ambient sound while the *mid (M)* pickup mostly picks up the direct sound from the instrument. The two are then mixed together by way of a transformer matrix to reconstruct a stereo image. By varying the direct vs. ambient sounds patterns, the stereo image can be made to be closer or more distant without moving the mike.

Fig. 5-12. M-S stereo miking technique pattern.

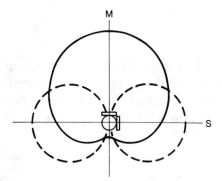

Microphone Choice and Placement

The most common miking situation in a modern multitrack studio will involve an instrument that is being recorded with the "close miked" method at a distance of one foot or less while using a single microphone. This mike output will then be processed by the recording console and sent to an individual track on the multitrack recorder. The studio may have more than one musician playing at one time but the above method would simply be duplicated each time, allowing for mike choice and placement differences among the instruments.

Charts 5-1 and 5-2 are presented as initial guidelines to mike choice and placement but, as this is a matter of individual taste, experimentation and experience may suggest additional approaches. Figs. 5-13 through 5-21 show several of the microphones used in a recording studio and their response curves.

Chart 5-1.
Common Miking
Techniques

DRUM SET

1. Two mikes placed over the drums, one on the left side and one on the right side to obtain a stereo effect.
2. Use a separate mike for each drum with the tom-tom mike set high enough to pick up the cymbals (eliminating the need for overheads) and the snare mike set high enough to pick up the hi hat.
3. Same as item 2, using a separate mike for the hi hat.

BASS DRUM

1. Remove the front head of the drum and place the mike inside the drum, off center. There are more overtones to the sides of the drum than in the center.
2. Place the mike off center at the rear (foot pedal) side, being careful that any squeaks from the pedal are not picked up.

SNARE DRUM

1. Aim one mike at the top head from a side angle.
2. Mike the top and bottom heads with separate mikes. The bottom mike gives the snare sound extra snap.

ACOUSTIC GUITAR

1. Place the mike behind the bridge.
2. Place the mike over the neck where it joins the body of the guitar.
3. Aim the mike into the hole of the guitar.

GRAND PIANO

1. Place the mike near the hammers at the center of the keyboard.
2. Place the mike at the center of the piano where the high and middle register strings are attached.
3. Place the mike at the rear of the piano where the bass strings are attached.
4. Hold the piano lid open using the long stick and place the mike at a distance from the piano aiming at the bottom of the lid.
5. Remove the lid from the piano and suspend the mike directly over the piano center.

UPRIGHT PIANO

1. Place a mike behind the sound board.
2. Place a mike over the open top of the piano.
3. Place a mike inside the piano.

HORNS

1. Place the mike close to the bell watching out for wind noise.
2. Place the mike two to three feet from the bell for a fuller sound.

VIOLIN AND VIOLA

1. Place the mike several feet above the instrument.
2. Place the mike close to the instrument aiming at the F-holes for a scratchier ''fiddle'' sound.

CELLO AND BASS FIDDLE

1. Place the mike over the bridge for a brighter sound.
2. Aim the mike into the F-hole for a fuller sound.

ELECTRONIC AMPLIFIER

1. Aim the mike at the center of the speaker cone for a bright sound.
2. Mike the speaker off center for a fuller sound.
3. Place the mike on the floor at a distance from the amplifier for a fuller sound.

LESLIE SPEAKER

1. Aim a mike into the top louvres only.
2. Aim separate mikes into the top and bottom louvres.
3. Place the mike about six feet away at the height of the Leslie. The rotating speaker reduces phase cancellation due to reflection from the floor. Better sound is obtained by placing the mike on a side of the Leslie that has louvres rather than at the back.

**Chart 5-1 (cont).
Common Miking
Techniques**

VOCALIST

1. Placing a mike within one inch or less from the vocalist's mouth produces a "present" breathy sound.
2. Placing the mike a distance of one to six inches from the vocalist reduces breath noise and crackling from the vocalist's mouth.

VOCAL CHORUS

1. Place a mike six to eight feet from the chorus.
2. Divide the chorus into several small groups and mike each group separately at a distance of one or two feet.

OBOE, CLARINET, AND FLUTE

Place a mike six to twelve inches above and over the finger holes.

TYMPANI

Place a mike eight to twelve feet from the instrument.

PERCUSSION INSTRUMENTS

Place a mike two to six feet away.

ORCHESTRAL BELLS, XYLOPHONE, VIBES

Place a mike four to six feet above the keyboard.

**Chart 5-2.
Common Mikes
Used for
Instruments**

Piano: U87, U89, C451, C414, CM1051, PZM, KM84.
Banjo, Guitar, Dobro: U87, U89, C451, C414, CM1051.
Bass Drum: U87, U89, D12, RE20, SM58.
Snare Drum: C451, MD421, KM84, SM57.
Other Drums: MD421, SM57, SM58, U87, U89, C451, PZM.
Horns: U87, U89, 77DX, SM57, U47, C414.
Strings: U87, U89, C451, C414, CM1051, PZM.
Amplifiers: U87, U89, RE20, D12, SM57, SM58, MD421.
Percussion: U87, U89, CM1051, C451, C414.
Vocals: U87, U89, U47, The AKG Tube, SM58, C414, RE2.

The above microphones are made by the following manufacturers:

AKG: C414, C451, D12, The AKG Tube.
CALREC: CM1051.
Crown: PZM.
Neumann: U87, U89, U47, KM84.
RCA: DX77.
Sennheiser: MD421.
Shure: SM57, SM58.

Fig. 5-13. The Neumann U87 condenser microphone *(Courtesy Gotham Audio Corp.).*

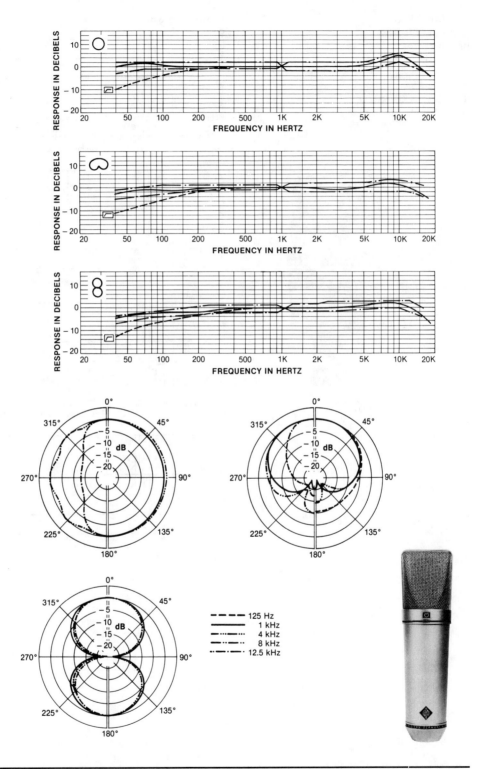

**Fig. 5-14. The
AKG 414EB-P48
pressure
gradient
condenser
microphone**
*(Courtesy AKG
Acoustics, Inc.).*

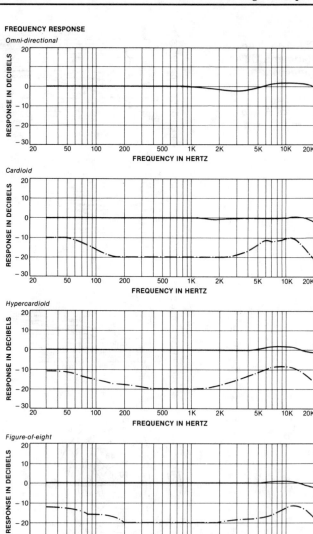

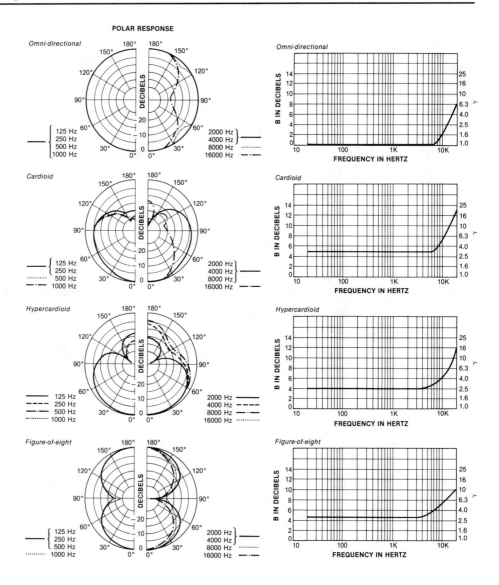

POLAR RESPONSE

Fig. 5-15. The C-460B pressure gradient condenser microphone *(Courtesy AKG Acoustics, Inc.).*

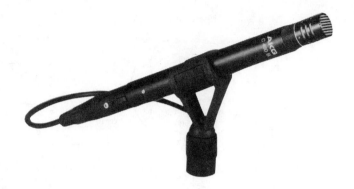

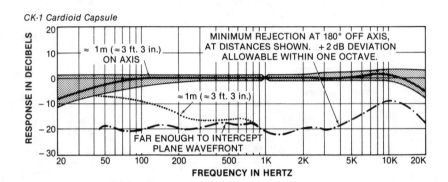

CK-1 Cardioid Capsule

≈ 1m (≈3 ft. 3 in.) ON AXIS

MINIMUM REJECTION AT 180° OFF AXIS, AT DISTANCES SHOWN. +2 dB DEVIATION ALLOWABLE WITHIN ONE OCTAVE.

≈1m (≈3 ft. 3 in.)

FAR ENOUGH TO INTERCEPT PLANE WAVEFRONT

RESPONSE IN DECIBELS

FREQUENCY IN HERTZ

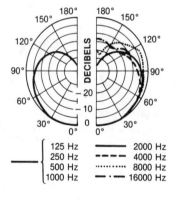

	125 Hz	2000 Hz
	250 Hz	4000 Hz
	500 Hz	8000 Hz
	1000 Hz	16000 Hz

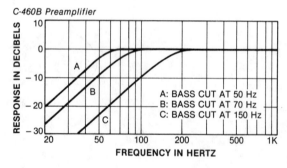

C-460B Preamplifier

A: BASS CUT AT 50 Hz
B: BASS CUT AT 70 Hz
C: BASS CUT AT 150 Hz

RESPONSE IN DECIBELS

FREQUENCY IN HERTZ

Fig. 5-16. The AKG Tube *(Courtesy North American Phillips Corp.).*

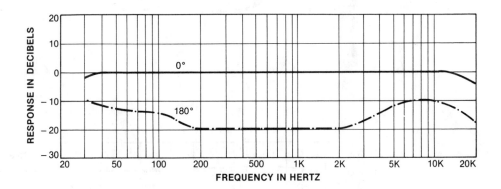

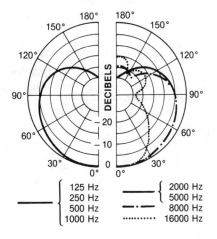

Fig. 5-17. The Calrec CM2056C windshield bass rolloff cardioid microphone *(Courtesy Audio Design Calrec Inc.).*

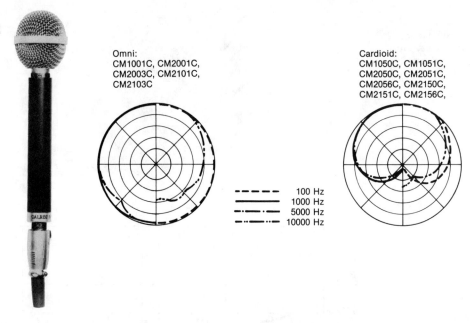

Omni:
CM1001C, CM2001C,
CM2003C, CM2101C,
CM2103C

Cardioid:
CM1050C, CM1051C,
CM2050C, CM2051C,
CM2056C, CM2150C,
CM2151C, CM2156C,

– – – – – 100 Hz
———— 1000 Hz
–·–·–·– 5000 Hz
–··–··–·· 10000 Hz

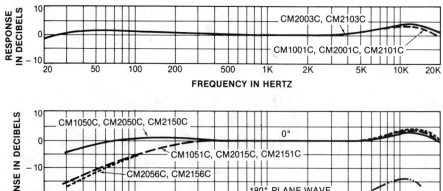

**Fig. 5-18. The
Beyer M 260
unidirectional
dynamic
microphone**
*(Courtesy Beyer
Dynamic, Inc.).*

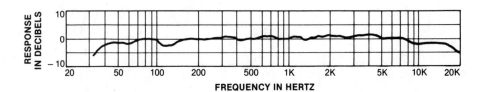

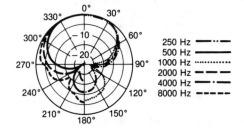

Fig. 5-19. The Electro-Voice RE-20 dynamic cardioid microphone *(Courtesy Electro-Voice Inc.).*

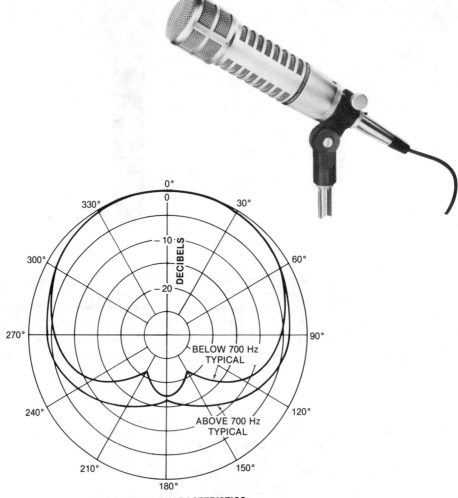

DIRECTIONAL CHARACTERISTICS

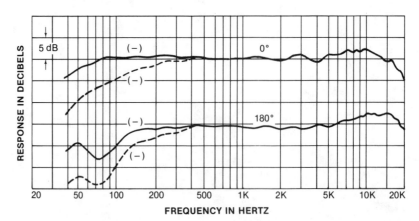

**Fig. 5-20. The
AKG D-12E
dynamic
cardioid
microphone**
*(Courtesy AKG
Acoustics, Inc.).*

FREQUENCY RESPONSE CURVES

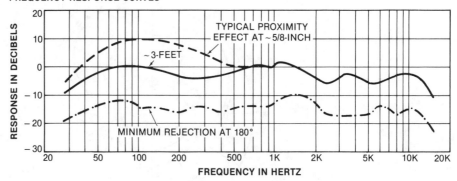

**Fig. 5-21. The
Shure SM57
dynamic
cardioid
microphone**
*(Courtesy Shure
Brothers, Inc.).*

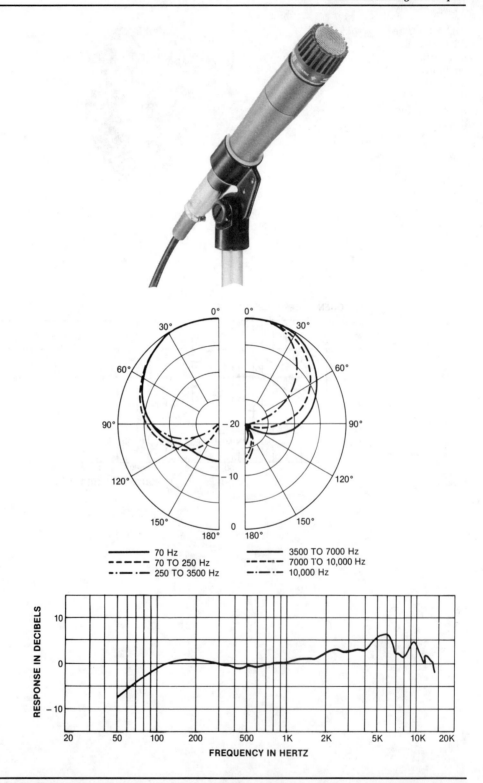

References

1. AKG, "M-S Stereo Recording Techniques," *Technical Manual,* North American Phillips Corp.

2. Anderson, Roger, and Robert Schulein, "A Distant Micing Technique," *db, The Sound Engineering Magazine,* Vol. 5, No. 4, April 1971, pp. 29-31.

3. Burroughs, Lou, *Microphones: Design and Applications,* Plainview, New York: Sagamore Publishing, 1974.

6 MAGNETIC TAPE RECORDING

Music recording in the modern studio, whether analog or digital, makes use of the magnetic tape recorder as the major storage device of music information. It is the function of any professional tape recorder to act as a memory device for the storage of a mass quantity of information and then, when required at any point in time, faithfully reproduce this information in its original form—music. The ideal recorder would be a memory bank of perfect accuracy with an unlimited capacity for storing separate groups of information in a constant time relationship. Magnetic tape is, at present, the most accepted and practical means of approaching this ideal.

The theory of tape recording is based on relating physical lengths of magnetic tape to periods of time (Fig. 6-1). By playing back each section of a length of tape at the same speed at which it was recorded, the original rhythm and duration of each sound and the spaces between sounds are preserved. The best way of ensuring that the time spectrum remains unchanged whenever the tape is played is to record and play it back at a constant speed. Although the methods by which information is stored are rapidly changing, the analog recorder has been the standard method of recording onto various kinds of material since its invention at around the turn of the century. In recent years, the digital revolution and the storage of information onto tape in the form of a digital stream of binary numbers has given new life to the recording industry by raising the clarity and quality of sound reproduction to new heights.

In this chapter, we shall discuss both digital and analog recording as well as the magnetic storage medium presently common to both.

Professional Audio Recorders

All professional reel-to-reel analog or digital audio recorders perform the same function, that of transporting magnetic tape across the

**Fig. 6-1.
Magnetic tape
recording
equates lengths
of tape to
periods of time.**

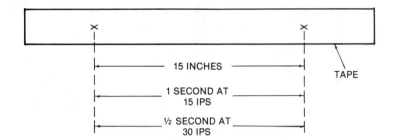

**Fig. 6-2. Studer
A80 master
recorder**
*(Courtesy Studer/
Revox of
America).*

recorder's tape heads at a constant speed under uniform tension. Shown in Fig. 6-2 is the Studer A80 ½-inch mastering machine transport. The features are as follows: (A) Supply reel, (B) Take-up reel, (C) Reel lock, (D) Capstan, (E) Capstan idler, (F) Headblock, (G) Tape guide and tension regulators.

Moving the tape at a constant speed is the main function of the *transport* which is actuated by pressing the *play button*. The *stop* button stops the motion of the tape by applying the brakes to the left and right turntables and by disengaging the constant-speed drive. The *fast forward* and *rewind buttons* engage the *tape lifters* to lift the tape off the heads and move the tape rapidly in either direction to find particular selections. Stopping a fast moving tape on a professional transport by simply pressing the stop button can be hazardous to the tape as there is a great deal of inertia due to the speed and weight involved, especially on multitrack tape reels. This could result in stretched tape or having tape all over the studio floor. To protect against this, "rocking" the tape to a stop is advised. This is accomplished by engaging the fast mode that is opposite to the fast wind direction of the tape until the tape shuttles to a slower speed, at which time, the stop button may be pressed. The *record button* enables signals to be recorded onto tape once the constant-speed drive mechanism has been engaged. Some recorders require that both the play and record buttons be depressed simultaneously in order to go into record, while others will drop in or out of the record mode by depressing the record button alone while in the play mode.

The *edit button*, commonly found on professional machines, has up to three modes of operation (consult the machine owner's manual): *stop-edit*, *play-edit*, and *fast-edit*. When the edit button is pushed while the transport is in the stop mode with the safety switch on, the left and right turntable brakes are partially released, and the safety switch is bypassed so that the tape may be moved easily from reel to reel by hand without the safety switch reapplying the brakes, if activated. If the edit button is pressed while the transport is in the play mode, the takeup turntable motor is disengaged, and the safety switch is again bypassed. This enables the tape to be played off the transport and into a wastebasket to remove unwanted material from a reel, while enabling the operator to hear what is being disposed of. Pressing the edit button in either the fast forward or rewind modes drops the tape lifters for as long as the button is pressed, in order to allow the operator to hear the tape at fast speed to find the beginning and/or end of selections.

The *safety switch* initiates the stop mode if activated by tape running out or breaking, except when defeated by the edit function. This switch may be incorporated into the tape tension arm, which helps take up slack when the tape motion is initiated. It may also be in the form of a light beam which simply senses the presence of tape in its path.

The *capstan* and *capstan idler* work together to move the tape. The capstan is rotated by a motor which turns at a constant speed. When the play button is pushed, the capstan idler presses the tape against the capstan, and capstan rotation pulls the tape past the heads. The take-up turntable motor is slightly energized so that it will turn and *take up* the tape that is pulled through the capstan. At the same time, the supply turntable exerts a force called *holdback tension,* which acts against the pull provided by the capstan. This provides tension on the tape as it passes the heads to assure good tape-to-head contact. The *inertia idler* (connected to a heavy flywheel) helps eliminate variations in tape speed due to friction (caused by the tape scraping against the supply-wheel flanges or being wound unevenly on the supply reel).

The transport system used by MCI, Ampex, Studer, and others is called an *open-loop system* and requires holdback tension to maintain tape-to-head contact. Another system used by some manufacturers, such as 3M, is called a *closed-loop* or *differential capstan* drive system. Figs. 6-3A and 6-3B show the locations of the drive parts: (A) Tape guide, (B) Incoming capstan idler, (C) Outgoing capstan idler, (D) Capstan, (E) Erase head, (F) Record head, (G) Reversing idler, (H) Playback head, (I) Tape lifter, (J) Head shield, and (K) Tape sensor lamp. In this system, the tape is pulled out of the head block faster than it is allowed to enter. The entrance and exit of the tape from the head block is controlled either by two separate capstans with different diameters or by one grooved capstan with two pressure rollers which mate with opposite grooves in the capstan (Fig. 6-4) so as to provide the same effect as two capstans with different diameters.

Fig. 6-3. The closed-loop tape drive system (dress covers removed) used by 3M in their Mincom Division tape machines *(Courtesy Mincom Division, 3M Co.).*

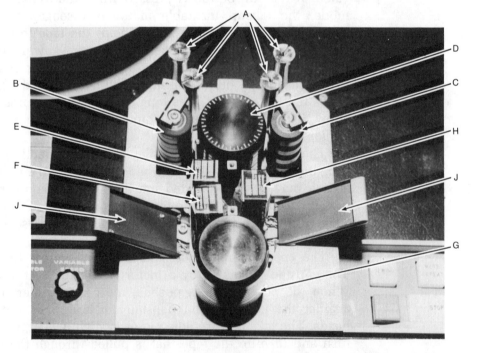

(A) Front view.

The phenomenon of removing tape from the head block faster than it is allowed to enter creates the necessary tension of tape against the heads and is possible due to the ability of the tape to be stretched by as much as 5% before permanent deformation takes place. The amount of stretch caused by closed-loop designs is well within this tolerance.

Another system is known as the zero-loop system and is becoming the most accepted design for both digital and the newer recorders. This system uses tension and motion sensors on both the supply and take-up sides of the head block (Fig. 6-5), which monitor the speed and tension of the tape, and applies corrective voltage changes to the supply and take-up turntable dc servo motors. This eliminates the need for capstans of any sort, making tape handling efficient and tape tension constant.

Scrape flutter, a problem which can plague analog recorders, is the vibration of a piece of tape in the direction of tape length due to its passing across the head. It is very much like the effect of bowing a violin string except that the bow (head) is stationary and the string (tape) moves across it. The shorter the unsupported tape length, the higher the flutter frequency until it is above audibility. To reduce the scrape flutter effect, *scrape flutter idlers* are often included on both

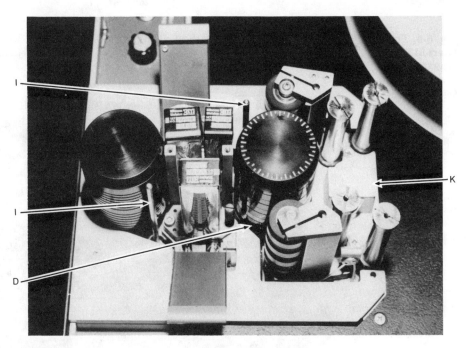

(B) Side view.

**Fig. 6-4.
Diagram of the
capstan and
capstan idlers
used in 3M's
"Isoloop" tape
drive. The
grooves make
the incoming
tape see a
smaller capstan
diameter than
the outgoing
tape** *(Courtesy
Mincom Division,
3M Co.).*

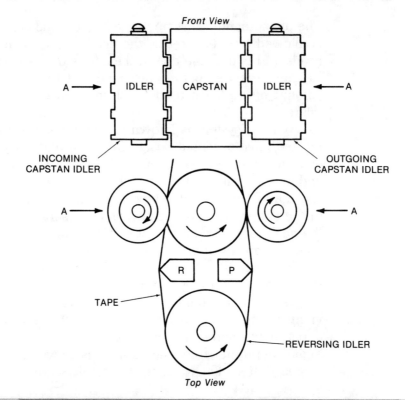

**Fig. 6-5. The
Otari MTR-90
tape transport**
*(Courtesy Otari
Corp.).*

sides of the record head to shorten the unsupported tape length to an acceptable value.

Almost all analog professional tape recorders have three tape heads: the erase head, the record head, and the playback head. The *erase head* wipes out any signals that were previously recorded on the section of tape passing it so that a new signal can be recorded. This allows recording on the same piece of tape many times. The erase head is only energized in the record mode. The *record head* converts the electrical energy of the signal fed to the recorder input into magnetic energy which can be stored on magnetic tape, while the *playback* or *reproduce (repro) head* converts the magnetic energy stored on the tape back into electrical energy which can be amplified and used to drive speakers. The head design of a digital stationary-head recorder incorporates an erase head and a record/playback head that are specially designed for digital reproduction.

If magnetic tape was a linear medium, the function of recorder electronics would be simple. Unfortunately, there is a discrepancy between the amount of magnetism applied to a piece of tape and the amount of magnetism retained by the tape after the magnetizing force is removed. The magnetism retained is called *remanence* and the greater the remanence of one tape with respect to another, the higher its output will be for the same applied magnetic recording force.

Magnetic Tape

Magnetic tape is composed of a *base material*, such as Polyvinyl-chloride (PVC), a strong durable material that can withstand a high degree of abuse before becoming permanently stretched or deformed. This base is used merely as a means of holding a magnetic-oxide coating and plays no direct role in the recording process. The molecules of the magnetic oxide form regions called *domains*. These domains are the smallest known permanent magnets. On an unmagnetized tape (Fig. 6-6A), the domains are oriented in random directions so that the north and south poles cancel each other out, leaving an average magnetic force of zero at any point on the tape. The individual magnetic fields are always present, however. When the tape is recorded, the individual domains are oriented in a definite direction by the magnetism of the tape head and are lined up in such a manner that their magnetism combines to produce an average magnetic force at the surface of the tape (Fig. 6-6B).

Fig. 6-6. Fields of domains on unmagnetized and magnetized tape.

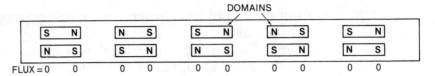

(A) Unmagnetized.

(B) Magnetized.

The Record Head

Fig. 6-7 is a diagram of a record head. Current flowing through the coils of wire wrapped around the *pole pieces* creates a magnetic force which flows through the pole pieces to the *gap*. Magnetic force, like electricity, flows easier through some mediums than others. In electricity, the amount of opposition to the flow of electric current is called *resistance*. The magnetic counterpart to current is *flux*, and the amount of opposition to the flow of flux is called *reluctance*. Since the magnetic oxide of the tape offers a lower reluctance path to the flux than the nonmagnetic material in the gap between the pole pieces, the flux flows from one pole piece through the tape to the other pole piece. The electrical signals fed to the record head are alternating currents and are, therefore, constantly changing in amplitude. As a result, the flux produced by the head is constantly changing. Since the recording gap has a certain length (measured in the direction of tape travel), a small section of tape can be magnetized to several different polarities and intensities as it passes the gap. The tape retains the last magnetic polarity and intensity orientation it receives before it leaves the gap. For this reason, the actual recording is said to be done at the *trailing edge* of the record head, with respect to the motion of the tape.

The Playback Head

The playback head (Fig. 6-8) is constructed similarly to the record head. When the magnetic flux on the tape passes through the gap, it induces a changing magnetic flux in the pole pieces. This flux cuts through the coils and induces a current in them, which can then be amplified. The flux flowing through the pole pieces of the playback head is a function of the average state of magnetization of the tape

Fig. 6-7. A record head.

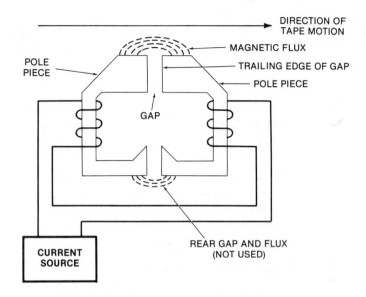

spanning the head gap at any given time. The electrical output of the head is proportional not only to the average magnitude of the flux, but also to its rate of change. If the tape was magnetized in only one direction with the same average magnetism at each point, there would be flux in the pole pieces but no output from the head windings. The rate of change of the magnetism on the tape increases as a direct function of the frequency of the recorded signal, so the playback head output is dependent on the recorded frequency even though the tape is magnetized to the same degree for each frequency. The output of the voltage of the head is proportional to

$$\frac{\Delta\phi}{\Delta t} \qquad \text{(Eq. 6-1)}$$

where,
   $\Delta\phi$ is a change in average value of gap flux,
   Δt is the time interval required for $\Delta\phi$.

Since the playback head output is directly proportional to the frequency recorded, the output of the head doubles for each doubling of the frequency on tape (Fig. 6-9). In terms of dB, this means that the output of the head rises 6 dB (2 times the voltage where dB = 20 log V_1/V_2) for each octave (2 times the frequency of the recorded signal).

**Fig. 6-8. A
playback head.**

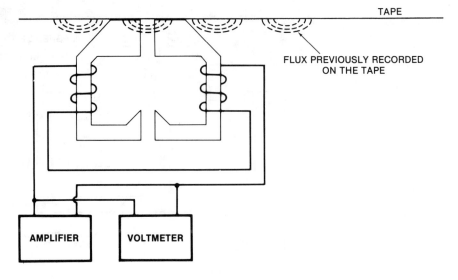

**Fig. 6-9. The
output of a
playback head
rises with
frequency until
the wavelength
of the recorded
signal
approaches the
gap length.**

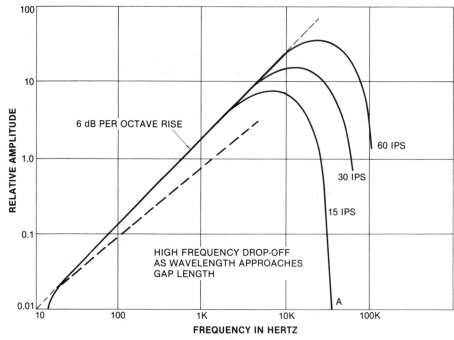

In addition to this change in output with frequency, the size of
the playback head gap is a factor in determining the frequency
response of the head. The head responds to the average value of magnetization in the gap, but as the frequency increases, more and more
of a complete cycle falls inside the gap at any point in time. Since a

sine wave has both positive and negative values, the true average magnetic flux of the gap decreases when more than half a cycle is spanned. This reduced output is caused by *scanning loss*. If the gap is spanned by exactly one cycle of a signal (i.e., a north magnetic field and a south magnetic field of equal intensity), the opposing fields cancel, and the average magnetization in the gap is zero, resulting in zero output from the head at that frequency (Fig. 6-10).

Fig. 6-10. The recorded wavelength is equal to the gap length, so the head produces no output.

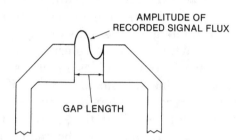

The upper frequency limit of the playback head is determined both by the length of the gap and by the speed of the tape. As the wavelength of the signal becomes less than twice the gap length, the output of the head decreases until there is no output at all when the wavelength equals the gap length. The wavelength of a signal on tape is equal to the speed of the tape divided by the frequency of the signal. For example, at 15 ips, one cycle of a 15-kHz tone takes up 0.001 inch of tape, while at 7½ ips, it takes up 0.0005 inch. Thus, the faster the tape speed, the higher the upper frequency limit of any playback head because the recorded wavelengths get longer at higher speeds. Similarly, the smaller the gap, the higher the frequencies that can be reproduced. Unfortunately, reducing the gap length to extend the high-frequency response has the drawback that average magnetism within the gap eventually decreases to the point where the output of the head is too low for a good signal-to-noise ratio.

The effect of too short a gap is also a problem at low frequencies. Below a certain gap length, low frequencies begin to roll off faster than 6 dB per octave since not only is the rate of change of flux falling due to frequency, but the amount of change that occurs within the gap is also decreasing because a smaller portion of a cycle is spanning the gap at any one time. Increasing the tape speed increases the wavelength of the signal and has the same effect on high and low frequencies as decreasing the gap length, because it is the ratio of the gap length to the signal wavelength that determines the head output. Thus, while 30-ips recording produces a more extended high-frequency response than 15 ips, a 30-ips response rolls off rapidly below 50 Hz, whereas response at 15 ips extends to 25 Hz. Other factors contributing to deviations from flat frequency response are electrical treble loss (due to hysteresis) and eddy currents (in the head core), which dissipate some of the magnetic flux in the gap as heat rather than converting it to electrical energy, and the capacitance between the head windings. These can result in a loss of 1 to 3 dB at 20 kHz. At low frequencies, a "contour effect" occurs that is due to the interaction of the entire head (rather than just the gap) with the long wavelengths of signals below a certain frequency. This effect, which is also referred to as "head bumps," produces an increase in output of as much as 3 dB, relative to what would be expected due to rate of change of flux. At 15 ips, these "bumps" cause a recorder aligned for flat response at 50 Hz to show an output of +1 dB to +3 dB in the 75- to 120-Hz range.

A compromise must be made between gap length, tape speed, and frequency response. Most present recorders have gaps that are between 0.00025 inch and 0.000038 inch wide.

Analog Tape Recording

The word analog means *similar* or *in direct relation with*. When applied to the analog tape recorder, it refers to the fact that the magnetic energy stored onto a magnetic tape, in the form of magnetic remanence, is in direct relation with and in proportion to the electrical waveform signal given at its input. An analog recorder requires special compensation circuitry in the form of equalization and bias current to operate within its linear limits with a minimum of noise and distortion. The following explanation and calibration sections shall refer to the analog recorder. The digital medium shall be treated separately, later in this chapter.

Equalization

To achieve a flat frequency response with magnetic tape, *equalization (EQ)* is used both in the recording and playback electronic circuits. Equalization is a term used to denote the changing of the relative amplitudes of different frequencies. The tone controls on a hi-fi system are an example of one type of EQ. Tape playback equalization boosts the low frequencies to compensate for the 6 dB per octave decrease of output as the reproduced frequency decreases (Fig. 6-11), while the record equalization boosts the high frequencies to compensate for the loss of some high-frequency energy by *self-demagnetization* and *bias erase*.

Fig. 6-11. The basic EQ used for magnetic recording.

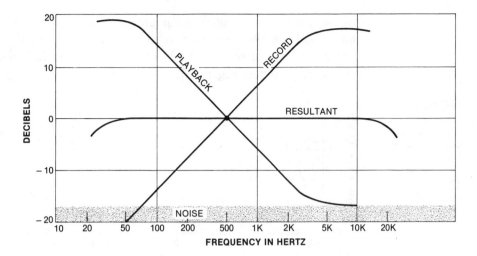

Fig. 6-11. The basic EQ used for magnetic recording.

Bias Erase

Bias erase occurs after the trailing edge of the record head and is the result of the formation of a secondary gap between the tape and the head as the tape leaves contact with the head. The flux generated in this gap by high-frequency signals and the bias current erases some of the recorded high frequencies the same way that the erase head erases a signal (Fig. 6-12). Since high frequencies tend to be recorded on the surface of the oxide coating, while low frequencies penetrate further into the oxide, the high frequencies are more susceptible to erasure. The bandwidth of the recording system is limited by the self-erasure of short wavelengths at the high end and by the signal-to-noise ratio at the low-frequency end. As the playback head output becomes very low, more amplification is necessary to bring it up to the same level as the higher frequencies, and, with enough amplification, the noise level of this playback amplifier becomes objectionable.

Fig. 6-12. High frequencies are partially erased by the flux induced in the bias erase gap by high frequencies and by the bias current.

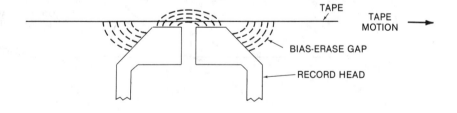

Bandwidth

The effective bandwidth for magnetic tape recording is about 10 octaves. The *bandwidth* refers to the frequency spectrum between the upper and lower *cutoff frequencies.* These are the frequencies on either side of the spectrum which are 3 dB lower than the center of the frequency spectrum; that is, they have one half the power of the other frequencies (Fig. 6-13).

Fig. 6-13. A frequency-response curve with a 20-Hz lower cutoff frequency and a 20-Hz upper cutoff frequency.

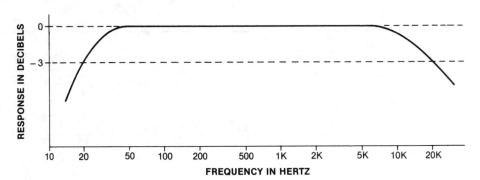

Bias Current

The magnetization curve of magnetic oxide is not a linear curve except between points A and B, and C and D, as shown in Fig. 6-14A; thus a method of reducing the distortion is needed. The method used involves mixing a certain amount of high-frequency current with the signal to be recorded. This current is called the *bias current* and must be several times the highest frequency to be recorded to prevent beats from being generated between it and harmonics of the input signal. In the Otari MTR-90 recorder, the bias frequency is 257 kHz; on other brand machines, the frequency will be different.

Fig. 6-14. Illustrating effect of biasing on recording linearity.

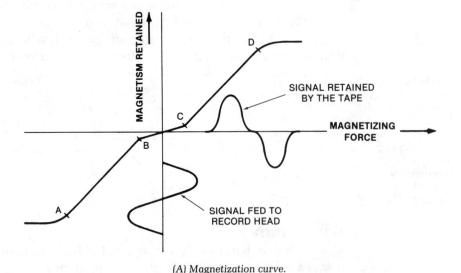

(A) Magnetization curve.

The bias current is mixed linearly with the audio signal so that there is no modulation of either frequency by the other. The bias signal moves the audio signal away from the nonlinear zero crossover point of the magnetization curve onto the two more linear portions of the curve (Fig. 6-14B). The bias signal itself becomes distorted by the magnetic properties of the tape, but since its wavelength is much shorter than the length of the playback head gap, it produces no output.

The amount of bias current used is very crucial and varies with individual record heads and different types of tape. The bias setting affects the sensitivity of tape both at high and low frequencies as well as affecting the overall frequency response, distortion, output level, and signal-to-noise ratio.

Erasure

The erase head is fed the bias current at a higher level than the record head so that the tape is saturated alternately in both the positive and negative directions. *Saturation* is the point where the tape is completely magnetized in one direction, so that additional increases in magnetic force do not cause additional increases in the magnetism retained by the tape. This alternate saturation destroys any magnetic pattern that might previously have been on the tape. As the tape moves away from the erase head, the intensity of the magnetic field decreases so that the degree to which the tape is magnetized in each direction also decreases until the domains are left in a random orientation, and the average magnetism is zero.

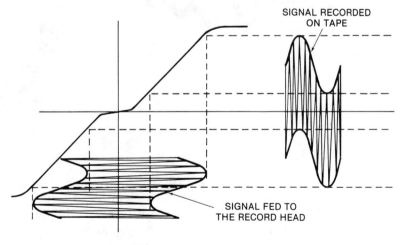

(B) After bias.

Recording Channels

Each channel of a multitrack tape recorder is identical electrically and in design to every other channel; they are simply duplicated by the number of channels provided by the machine. The recording channel electronics may be housed in a single complete module with the majority of the adjustment controls and switching functions designed right into the module's front panel or they may be set flush into the recorder's console area (Fig. 6-15). The electronics and adjustment controls may also be designed onto a single circuit board and then housed along with the other channels in a single mainframe. The mainframe is then fitted inside the recorder console out of view as shown in Fig. 6-16. This allows for ease in the servicing and interchangeability of the recording channel modules.

Fig. 6-15. One channel of electronics for an MM1000 or an AG440 with the plug-in bias, record, and reproduce amplifiers removed *(Courtesy Ampex Corp.).*

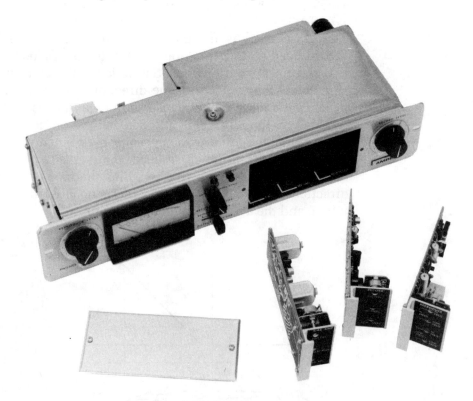

Professional recorders have three modes of operation: input, reproduction, and synchronization. In the *input mode*, the signal being monitored at the output and on the channel meters is the input signal, with the record/play signal being bypassed. In the *reproduce mode*, the signal being monitored at the output and on the channel

Fig. 6-16. Otari electronics in the console *(Courtesy Otari Corp.).*

meters is the playback signal—from the reproduce head. This allows you to monitor the entire record/playback chain, even while in the record mode. The remaining mode is the *sync mode* which allows the record head (on selected channels) to act as a playback head, thus keeping the music program in sync on a three-head recording system. In this mode, the signal being monitored is the signal output of the record head. The switches for these functions will often be located at the front of the machine and are also duplicated on a remote control unit that is located at the console position. Located with these functional switches are the *record enable* switches which are usually marked SAFE/READY. These record enable switches serve a safety function, preventing the accidental erasure or recording of a track unless physically engaged.

Located also on the channel modules are the level adjustments which allow for input level, output level, sync level, and low- and high-speed equalization trims. Refer to the operating manual for the machine for the location and typical settings of these adjustments.

Multitrack remote control units have, over the past decade, become quite sophisticated, most with all the necessary functions and track status controls located near at hand (Fig. 6-17). One useful feature built into most multitrack remotes is the *auto-locate* function. This function allows reference points placed on the tape to be stored into a memory. When called from memory by pressing the search button, the auto-locator will shuttle the tape to the requested point and either stop the transport or go directly into the play mode (if requested). Often, there will be multiple reference points which may be selected for various cues in a song.

Fig. 6-17. JH-110C-8 remote control with AutoLocator III *(Courtesy MCI/ Sony Corp.).*

Selective Synchronization

The use of *sel sync* (Ampex trade name for the sync or overdub position, and short for selective synchronization) arises from a need to hear previously recorded tracks while simultaneously being able to record another signal in sync with them on the same piece of tape. This process is called *overdubbing*. If sel sync is not used and this is attempted, the previously recorded signal is heard back by the repro head, while the new signal is recorded by the record head. The object of overdubbing is to be able to play these signals back together so that they sound like they were performed simultaneously. If sel sync is not used and the signals were now played back, the original track would be heard first with the overdubbed track lagging it by a time equal to the distance (d) between the record and playback heads, divided by the tape speed (Fig. 6-18). To prevent the overdubs from lagging the original tracks, the original channels are placed into the sel sync mode so that they are played back through their respective record head tracks. When the new signal is recorded, it is recorded directly above or below the basic tracks, and when the entire tape is played back through the playback head, the overdub is in sync with the basic tracks.

Fig. 6-18. If an overdub is attempted without the use of sel sync, the signal on track 2 will lag the original signal by a time equal to the distance (d), divided by the tape speed.

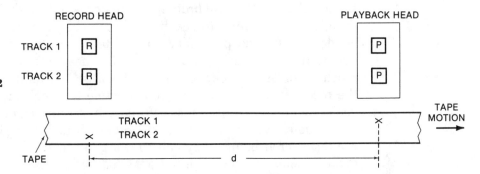

On most recorders, the frequency response of the signal played back in the sel sync mode is not as good as that played back through the repro head, especially in the low and high frequencies, because each head is optimized for its function. The signal is good enough, however, to be used to cue the overdubs and, on newer recorders, an additional equalization circuit has been added to improve the sel sync frequency response so that the difference between the record and reproduce heads is less noticeable.

Often, the recorder electronics will include a separate sel sync gain control, which enables the sel sync playback to be set to the same level as the normal playback so that no level changes need to be made when switching from one to the other. A bias trap adjustment is also often included. This control is set for each channel so that the bias signal being recorded on a track, which leaks into an adjacent head used for sync playback, is removed by a tuned LC circuit before reaching the playback channel output (Fig. 6-19). The bias leakage would be inaudible, but it can overload amplifiers following the recorder and, if not removed, would cause a false indication on the VU meter of the channel in sync.

Fig. 6-19. The sel sync bias trap.

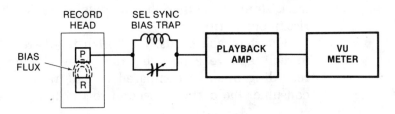

Calibration

Since the sensitivity, output level, bias requirements, and frequency response of different brands of tape vary considerably, the record level, playback level, bias current, record EQ, and playback EQ are all variable. The procedure used to set these controls is called *electronic alignment* or *calibration*. In major studios, this is done first thing in the morning for each machine. The alignment, however, is good only for one particular type of tape and one particular set of heads. If either the type of tape or the head block is changed, the machine must be realigned. As a check on the total performance of the recorder, however, it is good practice to check the alignment on each machine to be used just before each session. In this way, broken wires, defective components, and other obstacles to good recording are discovered before they can ruin a good take. The alignment procedure seems complicated, but once learned it usually takes less than ten minutes to complete.

Alignment is done with reference to an NAB (National Association of Broadcasters) standard playback alignment tape. This tape is available from Ampex, Standard Tape Laboratories, and others in ¼-inch, ½-inch, 1-inch, and 2-inch tape widths and is recorded *full track*. That is, the signal is recorded on the full width of the tape so that the signal is in phase at all of the gaps on an ideal multitrack playback head.

Head and Track Configurations

Several head and track configurations are available for each tape width. For ¼-inch tape, the most common ones are full track (called *monophonic*), two-channel, two-tracks (often called *half-track* or *half-track stereo*), and two-channel, four tracks (often called *quarter track* or *quarter-track stereo*). These are illustrated in Fig. 6-20.

Full- and half-track heads are used for professional work because the greater recorded track widths produced by them enable the tape to retain more magnetism and produce a higher output, resulting in better signal-to-noise ratios. The wide recorded tracks are also less sensitive to dropouts than narrow tracks. For this reason, ½-inch, ½-track stereo machines that operate at 30 ips have enjoyed a recent rise in popularity.

Quarter-track heads, for ¼-inch tape, are used in consumer reel-to-reel tape recorders because they double the available stereo recording time, as compared to half-track stereo heads, saving the consumer the cost of a second reel of tape. Two tracks are recorded

Fig. 6-20. Some track configurations available for magnetic tape recording.

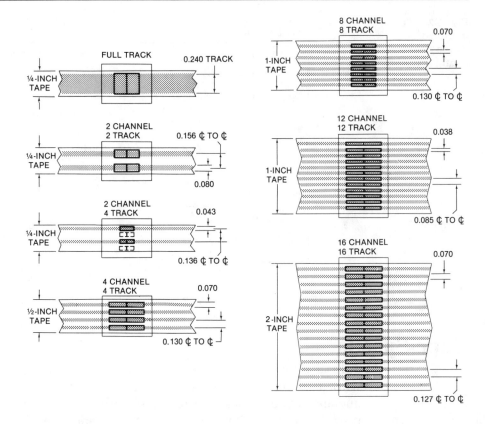

in one direction (Fig. 6-21A), the tape is flipped upside down, and recording is continued on the other two tracks with the tape moving in the opposite direction (Fig. 6-21B). If this were done with half-track stereo or full-track heads, the program recorded in the first pass of the tape would be erased during the second pass.

Since proper playback of the recorded information depends on using heads of the same configuration to record and play back the tape, studios use quarter-track heads to produce copies of tapes that clients can listen to on consumer-type tape machines. The original full- or half-track tapes are used to produce the discs or tapes that are sold in the record stores.

Guard bands of unrecorded tape are left between the tracks to prevent crosstalk. The width of the bands is only slightly less than the recorded track width, as can be seen from Fig. 6-20. Track widths are also available for recording half-track and four and eight channels on ½-inch tape, eight and sixteen channels on 1-inch tape, and sixteen and twenty-four tracks on 2-inch tape.

Fig. 6-21. The quarter-track stereo configuration.

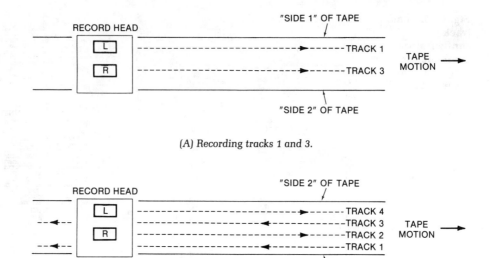

(A) Recording tracks 1 and 3.

(B) Recording tracks 2 and 4.

Tape Speeds

Several different tape speeds are used in the studio. A great deal of work is done at 15 ips because this speed allows all the audio frequencies to be recorded at full level without saturating the tape and it produces a good signal-to-noise ratio. In addition, this speed spreads out the recorded signal far enough apart for easy editing. Consumer hi-fi tape machines move tape at 7½ ips in order to save tape. Many ¼- and ½-inch studio recorders allow a choice of three speeds: 7½, 15, and 30 ips. The 7½-ips speed produces too much of a compromise in signal quality for use in recording multitrack master tapes, so 30 ips is most often available as a second multitrack speed. Since high frequencies are less likely to overload tape at 30 ips, a different record EQ, called AES (Audio Engineering Society) tape equalization, is used. This results in a signal-to-noise ratio at 30 ips which is better than 15 ips by 6 dB at 16 kHz, 3 dB at 8 kHz, and 2 dB at 4 and 2 kHz. At 1 kHz and below, the signal-to-noise ratio is the same as for 15 ips. The 30-ips speed also lessens the effect of tape dropouts because they last half as long; it increases the resolution of the recording by doubling the length of the tape, corresponding to a particular time interval, which results in better transient response; and it reduces high-frequency problems, such as playback-head scanning loss, azimuth-angle inaccuracies, and bias inaccuracies because these effects are wavelength dependent. Also, 30 ips doubles the wavelength of all frequencies with respect to 15 ips. Many state-of-the-art studios are using the 30-ips speed in conjunction with two-track ½-inch tape recorders to gain the further improvement in signal-to-noise ratio that is available by doubling the recorded track width. However, 30 ips does have some disadvantages:

1. At 15 ips, maximum print-through occurs at 1250 Hz, while at 30 ips, it occurs at 2500 Hz where the ear's better sensitivity can make it more annoying.
2. Tape costs are doubled and maximum recording time per 2500-foot reel of tape is 15 minutes. 3600-foot reels of tape are available which would increase recording time to 22 minutes, but their thinner backing material would increase print-through.
3. Low-frequency response rolls off below 50 Hz.
4. Low-frequency tape-modulation noise moves up an octave and becomes more audible.

Both 15 and 30 ips have become accepted present-day studio speeds. The combination of 30 ips with the use of low-noise high-output tape often makes noise-reduction devices unnecessary.

Playback alignment tapes are available in several different speeds and track configurations. Each speed has a different set of test tones recorded at specific levels and at specific frequencies. The 15-ips Ampex tape begins with a 700-Hz tone recorded at a *standard operating level* of 285 nanowebers/meter. All other tones are recorded in relation to this level so that when played back through a 15-ips NAB equalizer, all tones read the standard operating level. The NAB EQ compensates for the 6-dB-per-octave decrease in playback head output as frequency falls. The playback level control on each channel is set to produce the reference level at 700 Hz, which is 0 VU for use with regular tape and −6-dB VU for use with high-output low-noise tapes, such as Scotch 250 and Ampex Grandmaster. The playback level control is set 6 dB lower when recording on these tapes because they can retain *6 dB more signal* (Fig. 6-22A) before saturation occurs than regular tapes can. Setting the repro level for −6-dB VU with the alignment tape requires that the record level be boosted 6 dB during the record adjustments in order to obtain a 0-VU output level. The more playback gain used, the more tape noise is produced. Because the same output and distortion levels are obtainable from these tapes with less playback gain, a better signal-to-noise ratio results (Fig. 6-22B).

**Fig. 6-22.
Comparison of
high-output low-
noise tape with
regular low-
noise tape.**

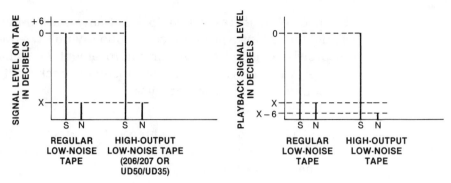

(A) Recorded signal level.   *(B) Playback signal level.*

After the 700-Hz tone are tones of 15, 12, 10, 7.5, 5, 2.5, and 1 kHz, and tones of 500, 250, 100, 50, and 30 Hz. The high-speed, high-frequency playback EQ is set to give the flattest high-frequency response (Fig. 6-23). This usually occurs when the high-frequency reproduce EQ is set so that the 10-kHz tone reads the reference level.

**Fig. 6-23. Close-
up of electronic
adjustments for
the Ampex
MM1100**
*(Courtesy Ampex
Corp.).*

The low-frequency playback EQ cannot be set using the standard full-track alignment tape due to an effect called *fringing*. Fringing occurs when a tape of one configuration is played back with a gap narrower than that which recorded the signal on the tape (Fig. 6-24). The longer wavelength signals (those below about 500 Hz), which are above and below the gap, are picked up by the head and added to those within the gap, producing an apparent rise in output at low frequencies. To avoid the fringing effect, the setting of the low-frequency playback EQ is postponed until after the record alignments are done.

Fig. 6-24. A half-track mono head reproducing a tape recorded full track.

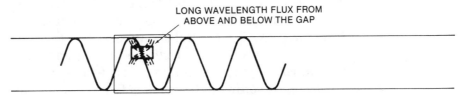

LONG WAVELENGTH FLUX FROM
ABOVE AND BELOW THE GAP

The 7½-ips Ampex playback alignment tape is similar to the 15-ips tape except that it lacks the 30-Hz tone and, also, all tones are recorded at −10-dB VU except for the last tone, which is a 700-Hz tone recorded at operating level. The lower level is necessary at 7½ ips to prevent saturation of the tape at high frequencies. Naturally, the low-speed, high-frequency playback EQ control is used at 7½ ips.

The alignment tape is allowed to run out in the play mode so that it is wound under constant tension to prevent it from stretching in storage. To save time, it is stored *tails out*, i.e., the end of the tape is on the outside of the reel, and it is not rewound to *heads out* until just before it is used again. The tape may be wound back and forth in fast forward and fast rewind to find the frequencies desired, but it must be played at its proper speed from start to finish before being stored.

Although a standard tape exists for setting reproduce levels and EQ for all NAB EQ tapes, the difference in sensitivities to high and low frequencies and the different bias requirements of different types of tape (even those made by the same manufacturer) require that record adjustments be made using the same type of tape that is to be used for the recording session.

The first step in the record alignment is the adjustment of the *erase adjust* control on the bias amp card. This control is used to tune the erase head circuit so that the bias current is a pure sine wave. If this control were not adjusted properly, the bias would be a distorted sine wave and would cause second-harmonic distortion of the recorded signal (frequencies which are twice that of the input signal would appear on the tape even though they are not present in the input signal). In addition, this bias distortion would cause the record head to become magnetized, erasing high frequencies and adding noise to the tape.

The erase peak adjustment is very stable and does not have to be made more than once a month unless the head block on the machine is changed. The erase adjustment can be made either with or without tape on the machine and with the safety switch defeated. All channels are set in the record mode and all output selector switches are in the bias position. The erase peak controls are then set for the highest readings on the individual VU meters. The reading will reach a peak, remain there as the control continues to be turned, and then fall back from the peak. The control should be set so that it is halfway between the point where the meter stopped rising and started falling. This is the proper setting for minimum bias distortion.

The next setting to be made is that of the amount of bias mixed in with the record signal through the BIAS ADJUST CONTROL. The type of tape to be used is threaded on the machine and a signal is fed in at operating level (0 VU on the console) at 1 kHz or 500 Hz, depending upon whether you are setting for 15/30 ips or 7½ ips, respectively. The output selector is placed in the reproduce position and the machine is set in the record mode. Since the amount of bias used affects the signal-to-noise ratio, distortion content, signal output, and frequency response of the recording, all these factors should be optimized when adjusting bias. As bias is increased to a certain point, all these factors improve. As bias increases past this point, low-frequency sensitivity, signal-to-noise ratio, and distortion content figures continue to improve, but high-frequency sensitivity begins to fall, thus reducing output at high frequencies. As bias continues to increase, signal-to-noise ratio and distortion figures improve further, but both low- and high-frequency sensitivity decrease so that the overall output is lower. Continuing in this same direction, the signal-to-noise ratio eventually begins to deteriorate because of the great amount of amplification needed to recover the ever-smaller signal recorded on the tape. The best compromise occurs near the level of bias which gives peak output of the signals in the 500- to 1000-Hz range. A little high-frequency sensitivity is sacrificed in favor of better signal-to-noise ratio and distortion figures, for this loss of sensitivity can be overcome with the record pre-emphasis (EQ).

The bias level is set by turning the control clockwise (increasing the bias) until the output of the tape rises to a peak. The bias level is then increased further until the output level falls 1 dB. This is called overbiasing the tape by 1 dB and with most low-noise, high-output tapes produces the lowest tape modulation noise. Overbiasing also decreases the sensitivity of the tape to *dropouts* which will be discussed later in this chapter. With some types of tapes, however, trying to restore the lost high-frequency sensitivity by boosting the record pre-emphasis causes a hump in the frequency response curve at around 6 to 8 kHz. For these tapes, the bias can be set for a lower amount of overbias but never less than the amount that produced the peak in the 500- to 1000-Hz range.

Some recorders offer a bias cal control which enables a reference level to be set for the bias in order to check it easily. After bias is set to the proper level, as described above, the output selector switch is set to *bias*, and the *bias cal* control is adjusted so that the VU meter reads 0 VU. Once adjusted, whenever the meter reads 0 VU in the bias position of the output selector switch, the bias is set properly. Since the proper bias control setting varies with the electrical characteristics of the heads, this setting is good only for one head block. This control is very stable and needs to be checked only once a month.

After setting the bias, the record level and pre-emphasis controls must be set so that all tapes recorded will play back according to the NAB standards. This assures interchangeability between studios. The gain and EQ of the playback head were previously set for flat frequency response when reproducing the NAB standard tape, so if the record circuits of the machine are now adjusted, the recorded signals having a flat frequency response when played back will produce a flat frequency response on all NAB standard machines.

First, a 700- or 1000-Hz tone is recorded on tape (the same type of tape for which we have just set the bias) with the output selector switch in the repro position, so that what has been recorded on tape and played back through NAB reproduce EQ can be monitored. The record-level control is adjusted so that the recorded signal plays back at a level of 0 VU. Next, the output selector switch is set to the input position, and the record calibrate control is set so that the meter again reads 0 VU. This equalizes the gain in the *input* and *repro* positions of the output selector so that the signal fed the machine and the playback of the signal being recorded on the tape can be compared at equal volumes to check its quality, and so that the engineer can see the level at which the signal will be recorded without actually having to record the signal.

The next adjustment is the record pre-emphasis control. This is set by feeding in a high-frequency tone, recording it on tape, and reading its playback level on the VU meter (output selector in the repro position). The control for the appropriate tape speed is set for the flattest high-frequency response. On most recorders, this occurs when 10 kHz reads the same level as the 700- or 1000-Hz reference tone. At 15 and 30 ips, this level is 0 VU, but at 7½ ips, the high-frequency record adjustment must be made at a record level of 10 dB below 0 VU. This is due to the large amount of EQ necessary to record high frequencies properly at low speeds. If the level were not reduced, the EQ would boost the high-frequency signals to the point that the tape would saturate. Once the tape reaches saturation, additional increases of magnetizing force causes no increase in the signal recorded on the tape. Thus, a 10-kHz tone cannot be recorded at operating level at 7½ ips. This does not affect music reproduction very much because the high frequencies are usually a very small proportion of instrument output.

After the record EQ is set, the low-frequency playback EQ can be set. This level could not be set with the full-track alignment tape because of the fringing effects discussed before. If we record a low-frequency tone now, however, it can be played back with the same head configuration that was used to record it, eliminating the fringing effect. A 50-Hz tone is recorded at operating level, and the low-frequency playback EQ for the appropriate tape speed is set so that the tone reads 0 VU on the meter with the output selector switch set on repro. This will now complete the record alignment procedure.

A step-by-step procedure for different speeds and different tapes is given in Chart 6-1, while the NAB curves for record and playback EQ at different speeds are given in Fig. 6-25; this also shows the CCIR record and playback curves. These curves are the European tape equalization standard and are used instead of the NAB standard in many places, although most major multitrack studios worldwide utilize the NAB standard.

Print-Through

After a recording is made, the engineer must take steps to prevent the signals stored on the tape from being altered inadvertently. One type of deterioration which can occur is called *print-through*. Print-through is the unwanted transfer of a signal from one layer of tape to another through magnetic induction. Its effect is more pronounced when excessively high levels are recorded and is increased when the tape is stored in or exposed to a magnetic field. The transfer from layer to layer decreases by about 2 dB for each 1-dB decrease in the record level but increases with the time of storage and increased

**Fig. 6-25. Pre-
and post-
equalization for
the NAB and
CCIR
characteristics.**

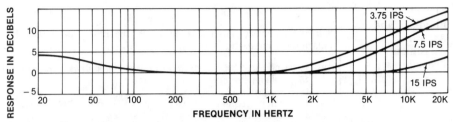

(A) Typical recording pre-equalization for ¼-inch tape recorders using NAB characteristics.

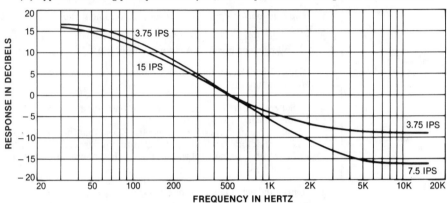

(B) Typical post-equalization for ¼-inch tape recorders using NAB characteristics.

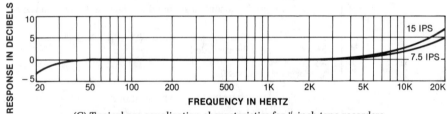

(C) Typical pre-equalization characteristics for ¼-inch tape recorders
running at 7.5 and 15 ips, using CCIR (DIN) standard.

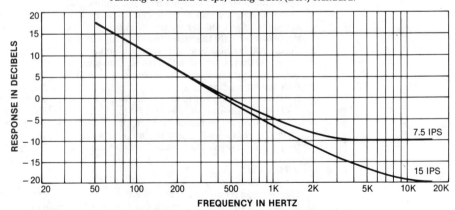

(D) Typical post-equalization curves for ¼-inch tape recorders
using CCIR characteristics at 7.5 and 15 ips.

**Chart 6-1.
Playback and
Record
Alignment
Procedures**

PLAYBACK ALIGNMENT

Thread the playback alignment tape on the machine to be aligned. If the tape was stored tails out, rewind to the head.

A. *15-ips playback alignment for low-noise, high-output tape, using a standard full-track 15-ips alignment tape.*
   1. Set repro level for 0 VU at 700 Hz.
   2. Set high-frequency 15-ips playback EQ for 0 VU at 10 kHz.
   3. Reset repro level for − 6 VU at 700 Hz.
   4. Do not adjust low-frequency playback EQ until after record adjustments.

B. *15-ips playback alignment for regular tape using a standard full-track 15-ips alignment tape.*

   1. Follow steps given in A, above, but omit A3.

C. *7½-ips playback alignment for low-noise, high-output tape, using a standard full-track 7½-ips alignment tape.*

   1. Set repro level control so that the 700-Hz tone recorded 10 dB below operating level reads 0 VU.
   2. Set 7½-ips high-frequency playback EQ so that the 10-kHz tone reads 0 VU.
   3. Do not adjust low-frequency EQ until after record adjustments.
   4. Set repro level control so that the 700-Hz tone recorded at operating level reads − 6-dB VU.

D. *7½-ips playback alignment for regular tape, using a standard full-track 7½-ips alignment tape.*

   1. Follow steps C1 through C3.
   2. Set repro level control so that the 700-Hz tone recorded at operating level reads 0 VU.

RECORD ALIGNMENT

Thread the tape to be recorded on the machine.

A. *For all tapes at 15 ips.*

   1. Set output selector to the bias position, and set the machine into the record mode on all tracks.
   2. Adjust the erase peak control for maximum meter reading.
   3. Feed a 1000-Hz tone into the machine inputs.
   4. Set the output selector to repro and, beginning with a low bias setting, increase the amount of bias until the meter reading rises to a maximum. Continue increasing the bias level until the meter reading drops by 1 dB.
   5. Feed a 700-Hz tone into all machine inputs at a 0-dB level from the console and set the level control so that the meter on the recorder reads 0 VU.
   6. Set the output selector to the input position and adjust the record cal control for a 0 VU meter reading.
   7. Set the output selector to repro and feed a 10-kHz tone into all machine inputs at 0-VU level from the console, and set the 15-ips record EQ so that the meter reads 0 VU (at the peak of the meter swing if the needle is unsteady).
   8. Feed a 50-Hz tone into the machine inputs at 0-VU level. Adjust the 15-ips low-frequency playback EQ for a 0-VU meter reading.

B. *For all tapes at 7½ ips.*

   1. Follow steps A1 and A2 above.
   2. Set the output selector to repro and feed a 500-Hz tone to all the machine inputs. Set all tracks into the record mode and increase the bias level to obtain maximum

Chart 6-1 (cont).
Playback and
Record
Alignment
Procedures

meter reading. Increase bias so that meter reading drops slightly below the peak reading.

3. Feed a 700-Hz tone at 0 VU to all machine inputs and set the level control for a 0-VU reading on the meter.

4. Set the output selector to input and set the record cal control for a 0-VU reading on the meter.

5. Reduce the setting of the record level control so that the meter reads −10-dB VU.

6. Set the output selector to repro and adjust the repro level control for a 0-VU reading on the meter.

7. Feed a 10-kHz tone into the machine inputs at 0 VU and adjust the 7½-ips record EQ to obtain a meter reading of 0 VU.

8. Feed a 50-Hz tone into the machine inputs at 0 VU and adjust the 7½-ips low-frequency playback EQ for a 0-VU meter reading.

9. Feed a 700-Hz tone into the machine inputs at 0 VU, set the output selector switch to input, and adjust the record level control for a reading of 0 VU on the meter.

10. Reduce the repro level control to approximately the position set with the operating level tone on the playback alignment tape. Then switch the output selector to the repro position and adjust the repro level control for a 0-VU meter reading. (Follow step 10 in the order stated to prevent pinning and possibly damaging the VU meters on the tape machine.)

storage temperature. The amount of print-through depends upon the physical separation of the section (carrying the printing signal) from the tape that is to be printed on, so tapes with thinner backings and with high-output oxide tend to print-through more.

Magnetization components are recorded both in the direction of tape length and perpendicular to its surface. According to E. D. Daniel,[2] these components print onto the adjacent layers of tape where they create external fields which are in phase at the surface of the outer layer and are out of phase at the surface of the inner layer. In playback, the fields add for the outer layer and partially cancel for the inner layer (the fields are of unequal strengths), producing less audible print-through for the inner layer than for the outer (Fig. 6-26).

Fig. 6-26. The difference between print-through to the inner and outer layers of tape.

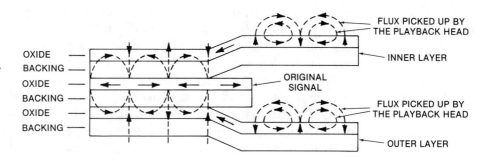

If the tape is stored heads out, the print-through will be heard before the main signal occurs. If the tape is stored tails out, the louder print will come after the signal occurs and will be masked somewhat by the desired signal. Print-through is not always permanent. By storing tape tails out and rewinding just before playback, the printed signal intensity is reduced, and there is usually not enough time for a new signal to be printed before the tape is played back. Storing tape outside the temperature range of 60 to 80 °F, or outside the range of 40 to 60% relative humidity, can shorten the life of both virgin and recorded tapes due to stretching and contracting of the base material.

Degaussing

Tape recorders must be clean both physically and magnetically to give optimum performance. If parts of the machine over which the tape passes are magnetized, erasure of the high frequencies can occur. If the record head is magnetized, the signal on the tape will have second-harmonic distortion.

Magnetism can be removed by the use of a *degausser* or *demagnetizer*. The degausser acts in the same way that the erase head of a recorder works, except that it operates at the power-line frequency of 50 or 60 Hz instead of above 100 kHz. This lower frequency is effective only if the degausser is moved past the magnetized object very slowly. The degausser works by subjecting the magnetized object to a magnetic field of sufficient strength to overcome that already present in the object. Once that occurs, the alternating field causes the domains of the object to be alternately magnetized one way and then the other. If the magnetic field is then slowly decreased, the domains are left in a random pattern such that the overall magnetism is zero. This is accomplished by moving the degausser away from the object at a speed no greater than one or two inches per second. The degausser must not be energized or de-energized within three feet of the tape recorder, or flux surges may create magnetic charges larger than that which the degausser can remove. Each object in the tape path must be slowly approached, saturated magnetically, and slowly moved away from, one by one, for proper demagnetization.

A *magnetometer* or *gaussmeter* is a device that measures the magnetic charge on an object. It is useful for indicating when an object is charged and needs degaussing, but since it is sensitive to the average of the magnetic charges over a rather large area, the object being measured may be charged even though it causes no reading on the meter. For example, the two pole pieces of a playback head could be magnetically charged, one north and one south. Since the sensitive part of the magnetometer covers both pole pieces at once, the fields cancel, and it gives no reading. The tape passing over the playback head, however, would receive the charges individually and possibly be partially erased. For this reason, degaussing is done every morning.

Cleanliness

The tape recorder also accumulates dirt due to *oxide shed*. This is oxide which falls off of the magnetic tape and can be accumulated on the heads, guides, and pinch roller. If it is allowed to build up on the heads, it can cause *dropouts*, which are momentary drops in the amplitude of a signal. This is due to the *separation loss* which occurs when the tape moves away from the *intimate* or *close* contact with the tape heads that is provided by the holdback tension of the supply reel. This loss can be computed as follows: separation loss (in dB) equals 55 times the separation distance, divided by the wavelength of the signal being recorded or reproduced. A separation distance equal to the thickness of a piece of cellophane from a cigarette pack (0.001 inch) can cause a dropout of 55 dB at 15 kHz (the wavelength of a 15-kHz signal recorded at 15 ips is 0.001 inch). Thus, any dirt which pushes the tape away from the head can cause severe problems.

Similarly, if dirt is allowed to accumulate on guides or on the pressure roller where it might be transferred to the tape, causing the tape to push away from the head, dropouts can occur. The machines should always be inspected for cleanliness before recording or playing back tapes. Any dirt, oxide shed, or pieces of tape that result from editing or tape breakage should be removed before beginning a session. Sensitivity to dropouts occurring during recording is decreased by overbiasing the tape, for as the dropout occurs and the oxide coating moves away from the head, the bias put on the tape decreases, causing the output of the tape to increase.

Head Alignment

Another very important factor affecting the quality of a recording is the physical head alignment. The record and playback heads in the head block have five adjustments: height, zenith, wrap, rack, and azimuth.

The *height* determines where across the width of the tape the signal will be recorded (Fig. 6-27). If the track is recorded and reproduced on heads with different height settings, not all of the recorded signal will be reproduced, resulting in a poorer signal-to-noise ratio, and if the tape is multitrack, crosstalk will occur between the tracks.

Zenith refers to the tilt of the head towards and away from the tape. The zenith must be adjusted so that the tape contacts the top and bottom of the head with the same force, otherwise the tape will tend to *skew*.

**Fig. 6-27. The
head gap width
must be centered
on the track
location.**

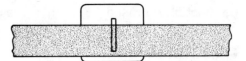

**Fig. 6-27. The
head gap width
must be centered
on the track
location.**

Skewing is the riding up or down of a piece of tape on a head or
guide, so that its edges are no longer parallel to the top plate of the
transport. This causes variation in the effective height, azimuth, and
tape speed.

Wrap refers to the angle at which the tape bends around the head
and the location of the gap in the angle. The wrap determines the
intimacy of the tape-to-head contact and thus controls the sensitivity
of the head to dropouts.

Rack refers to how far forward the head is and determines the
pressure of the tape against the head. The farther forward the head
is, the greater the pressure.

Azimuth refers to the tilt of the head in the plane parallel to the
tape motion (Fig. 6-28). The head gaps should be perpendicular to the
tape, so that all of the gaps are in phase with each other and so that
the signal is in phase with all points within each gap.

**Fig. 6-28. The
head gap width
must be set
perpendicular to
the tape edge.**

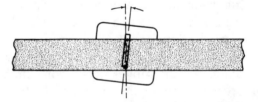

Height can be adjusted optically or by using a test tape of the
proper track configuration and adjusting the head height for maximum output at 1 or 3 kHz. Zenith is tested by covering the pole
pieces with a white grease pencil and playing a piece of tape to
observe the pattern formed as the grease pencil wears off (Fig. 6-29).
Naturally, this tape cannot be used for recording afterwards since
the grease pencil rubs off onto the oxide. The edges of the wear pattern should be parallel. If they are not, the zenith must be adjusted
using the screws on the head block.

Wrap angle is checked at the same time as zenith by making sure
that the wear pattern is centered around the gap. Need for rack
adjustment is indicated if the grease pencil wear pattern is wider on
the record head than on the playback head, or vice versa. Azimuth

Fig. 6-29. Grease pencil wear patterns for checking zenith adjustment.

GREASE PENCIL NOT WORN OFF HERE

(A) Correct.   (B) Bottom of head too far forward.   (C) Top of head too far forward.

can be tested by deliberately skewing the tape across the heads (by pushing up and down on the edges of the tape right in front of the head), while reproducing the 15-kHz band of a standard alignment tape. If the output increases, the azimuth needs to be adjusted. It can be adjusted by either of two methods, both of which use a full-track standard alignment tape.

The first method for adjusting azimuth is to play the 15-kHz section of a standard alignment tape and adjust the azimuth for the highest output on all channels. This will have to be a compromise because some channels will rise as others fall on either side of the correct setting. The peak at the correct setting is fairly sharp with smaller broader peaks to either side of it. In order to be sure of finding the proper peak, the head gap should be perpendicular to the tape path before the adjustment is attempted.

The second method uses the phase of the 12- or 15-kHz signal to find the correct setting. After finding the peak, as in the first method, the output of the top channel of the head is fed into the vertical input of an oscilloscope, and the bottom channel is fed into the horizontal input of the scope. The pattern resulting on the face of the scope represents the relative phase of the two channels. A straight line sloping up 45° to the right indicates that the two channels are in phase, a circle indicates that they are 90° out of phase, and a straight line sloping upwards 45° to the left indicates that the two channels are 180° out of phase. The azimuth should be adjusted so that the two channels are in phase.

On a multitrack head, it is not possible to get all the gaps in phase at the same time. This is due to the *gap scatter* which occurs in manufacturing, so the phase adjustment is made for the outer tracks. The inner tracks will then be usually less than 60° out of phase with one another. The record-head azimuth can be set in the same way as the playback-head azimuth by playing the test tape back in the sync mode. Erase-head azimuth is not very critical because its gap is very wide. It is correct as long as it is approximately at right angles to the tape path.

Head adjustment need not be done too frequently. Its necessity is indicated by the deterioration of performance of the tape machine as indicated by the inability to align the electronics to their normal response characteristics. It should be checked periodically by the studio technicians.

Figs. 6-30 through 6-37 are some of the tape machines currently in use in recording studios.

**Fig. 6-30.
Ampex AG440
2-track and 4-
track recorders**
(*Courtesy Ampex
Corp.*).

**Fig. 6-31. The
Ampex MM1100
16-track
recorder**
*(Courtesy Ampex
Corp.).*

**Fig. 6-32. The
3M M79/24, a
24-track
recorder. This
machine is also
available in 8-
and 16-track
versions**
*(Courtesy
Mincom Division,
3M Co.).*

Fig. 6-33. The MCI JH-24-24 recorder with remote control and AutoLocator *(Courtesy MCI, a Division of Sony Corp. of America).*

Fig. 6-34. The Otari Model MTR-90-II Channel Master recorder *(Courtesy Otari Corp.).*

Fig. 6-35. The Otari MTR-12 ¼-inch two-channel or ½-inch four-channel mastering recorder *(Courtesy Otari Corp.).*

Fig. 6-36. The Studer A800 recorder *(Courtesy Studer/ Revox of America).*

**Fig. 6-37. The
Studer A820
analog master
recorder**
*(Courtesy Studer/
Revox of
America).*

Digital Tape Recording

As we have seen, conventional analog recording equipment records the original audio input signal waveform in the form of remanent magnetization of the recording tape. This means, of course, that the nonlinearities of the magnetic tape distort the waveform, along with wow-and-flutter, modulation noise, and other irregularities; this results in the degradation of the original signal at playback. With digital recording equipment, however, the input signal is converted into digits or groups of pulses, which are then recorded onto tape, with the end result being a recording which is free of all the nonlinearities inherent to the analog medium.

The basic concept of *digital* or *pulse-code modulation (PCM)* recording is relatively simple to understand in theory. All basic audio waveforms can be broken down into two components: amplitude and time, as shown with a sine wave (Fig. 38B). A digital system operates by separating time into very short segments, the length of which is determined by a clocking frequency called the *sampling rate*. The sampling rate, which nominally operates between 45- and 50-kHz, breaks the audio waveform down into minute, stepped segments (Fig. 38A) with each segment having a specific voltage level (at one instant in time) which is sampled by an analog-to-digital converter and assigned a specific binary stream of number bits (called a word). Note that each sample is measured and expressed as a *binary number*. This binary number consists of a series of "0" and "1" digits with the "1" being represented by a pulse and the "0" by the absence of a pulse. At each sampling point, the binary number constitutes a numeric representation of the recorded waveform signal. This digital stream of words may then be recorded onto any number of mediums that can store digital pulses, such as magnetic tape, compact laser disc, etc. Once stored in a chosen medium, the pulses may be reproduced simply by reading the stream of digital words at the precise sampling rate at which they were recorded. Each word is then read and reconverted back into its original voltage level by way of a digital-to-analog converter. The accuracy of playback is controlled precisely, by the crystal-locked sampling rate, which eliminates all effects of wow-and-flutter and modulation noise.

Fig. 6-38. Sine-wave output of a digital playback processor.

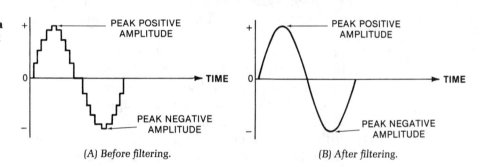

(A) Before filtering.   (B) After filtering.

Since it is the presence or absence of a pulse which contains the information, media-induced noise or changes to the shape of the pulse (due to nonlinearities of the media) are no longer important. As long as the difference between the existence or nonexistence of a pulse can be detected, the signal can be recovered accurately.

As digital processing works by sensing voltage levels at each sample pulse, the PCM system records and plays back discrete steps in voltages at each clock pulse. Thus, a digitally recorded signal would be reproduced as a series of very closely spaced level steps and not as a continuous waveform (Fig. 6-38A). To overcome this, a filter is introduced after the digital-to-analog conversion, which acts to smooth out the resulting steps and restore the waveform back to the analog equivalent (Fig. 6-38B). Some performance specifications for a typical PCM digital recording system are given in Chart 6-2.

Specifications

The signal-to-noise ratio of a PCM system is governed by the number of binary bits that are included in a digital word (present technology is based on a 16-bit word system), and by the noise figures of the

**Chart 6-2.
Performance
Specifications
for a PCM
Recorder**

Dynamic range	Better than 90 dB
Frequency response	10 Hz–20 kHz (± 0.5 dB)
Total distortion	0.02% (at peak level)
Crosstalk	− 85 dB (1 kHz)
Wow-and-flutter	Limited only by quartz crystal oscillator
Playback signal-level variation	None
Print-through	None
Residual level after erasure	None

analog-to-digital (A/D) and digital-to-analog (D/A) converters. Given a properly designed system, the signal-to-noise ratio for a signal coded with N bits is given by:

$$S/N = 6\,N + 1.8 \quad (dB)$$

(Eq. 6-2)

For a 16-bit system, this would yield a noise figure of 97.8 dB, a figure exceeding conventional analog recorder capability by about 30 dB.

Frequency response of a PCM system is governed by the level-detection circuit in the A/D and D/A converters. In a 16-bit system, the number of possible discrete level steps are 16^2 or 256 steps at any one point in time. Thus, when a signal is recorded and translated into a digital word, upon playback this word will be translated back into the same precise voltage level. It is important to note that non-uniformities in the tape, which affect the level or frequency response of analog signals, will not affect PCM performance as the tape is simply a storage medium for pulses representing binary digital words. PCM frequency-response performance is a function of electrical circuit design rather than the characteristics of magnetic tape.

The wow, flutter, and modulation noise associated with tape path irregularities are virtually eliminated with digital processing as the record/playback system is locked in time by the clocking frequency of the quartz crystal. Upon playback, the digital stream of words are first written into a buffer memory for temporary storage and then are read out in perfect synchronization with the clocking frequency, thus eliminating all effects of tape path irregularities.

Print-through and crosstalk are reduced to a negligible level due to the digital nature of the medium. Where there is a detectable signal pulse, it results in an accurate reproduction, and with a total absence of signal, it results in errors.

Error Correction

Information density, when using PCM recording and playback equipment, is generally extremely high (about 20,000 bits per inch at

15 ips on a fixed head deck), so that dust, etc., adhering to the surface of the magnetic recording medium will readily generate error signals. Error-correction codes are therefore normally added to the signal so that should an error arise, it will be automatically detected and corrected. If the errors are too numerous to be corrected this way, error-concealment techniques are brought into use. These measure the error-free signals immediately before and after the error sample and substitute their average value in place of the error sample. The effectiveness of these error-correction and error-concealment functions is critically important for the stable and reliable operation of PCM digital equipment. The three major error-correction/concealment codes are the Reed-Soloman code (an error-correcting code that works across the track widths of the tape), the cyclic-redundancy code (which detects errors along the width of each tape track), and the cross-interleave code (which reduces dropout errors).

Fixed- and Rotating-Head PCM Systems

PCM recording equipment may be divided into two professional system types:

1. Fixed head.
2. Rotating head.

The fixed-head PCM recorder is a reel-to-reel-based recorder, which often closely resembles an analog recorder in design and function. Mitsubishi Electric Corp. has developed two fixed-head PCM recorders, the X-80 (Fig. 6-39) and the X-800 (Fig. 6-40). The X-80 recorder is a two-track mastering PCM machine, whose format utilizes 10 tracks across a $\frac{1}{4}$-inch tape (Fig. 6-41). Eight PCM tracks, which are derived from the two audio program channels, are located across the width of the tape. A combination of the cyclic redundancy code (CRC), which detects errors along the tape's path length, and the Reed-Soloman Code, an error-correcting code detecting across the width of the tape, ensure that all errors in the signals coming from one or both tracks can be corrected. In addition to the digital tracks, an *analog cue* and *SMPTE* track are provided on the outside bands of the tape. Tape defects, arising from dust or dirt, will rarely extend over two track widths across a tape so that, under most conditions, nonrecoverable error signals will not arise. When errors beyond the capability of the error-correction system arise, the error is detected and dealt with in most PCM machines by *muting*, completely interrupting the program material. In the X-80 recorder, a *cross-fade* technique is employed which averages the signal both before and after the error point, smoothing out the splice transition.

Fig. 6-39. The Mitsubishi X-80 2-channel digital audio recorder *(Courtesy Mitsubishi Electric Corp.).*

Fig. 6-40. The Mitsubishi X-800 32-channel digital audio recorder *(Courtesy Mitsubishi Electric Corp.).*

Fig. 6-41. Track format of the Mitsubishi X-80 recorder. Analog tracks are recorded at each side of the tape for locating the editing point *(Courtesy Mitsubishi Electric Corp.).*

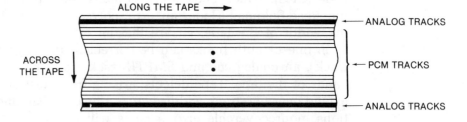

One feature of the X-80 recorder is that manual tape-splice editing is possible. This is not the case with most PCM machines for two reasons:

1. The synchronization of the PCM signal with the internal clock makes it very difficult to locate an editing point by "rocking."
2. Most PCM machines cannot digitally cross-fade at edit points, but will mute the signal in the presence of a large amount of errors.

The X-80 recorder provides an *analog cue track* which allows a manual rocking of the tape back and forth to find the *edit point*. This cue track may be listened to over the studio monitors or over an internal cue amp/speaker combination that is built directly into the recorder. Once the edit points have been found, a splice vertical to the tape is made by a special editing block and the two edit points are joined together.

The Mitsubishi X-800 (Fig. 6-40) is a fixed-head 32-track PCM recorder, which utilizes standard 1-inch videotape at an operating speed of 30 ips. All 32 channels are available at all times for recording and overdubbing; in addition, there are 3 digital tracks and 2 analog tracks that are available for SMPTE or information storage. With a high-density data storage that is typically greater than 30,000-bits per inch, a comprehensive error-correction system has been developed to correct for errors which will inevitably arise. Mitsubishi Electric Corp. has devised what they term as a semi-separate error-correction format. In this system, one audio channel is assigned to one digital PCM track on the tape. For error correction, two parity check tracks are provided for each subgroup of 8 channels. This requires 40 digital tape tracks in order to record 32 audio channels. In addition, each channel has its own cyclic redundancy code for detecting errors along the tape's width, which gives extensive overall error correction.

Sony Corp. has also developed a 2-channel multitrack digital recorder using the Digital Audio Stationary Head (DASH) format, a format jointly announced between Sony, Willi Studer, Matsushita Electric Industries, and MCI. The PCM-3102 2-channel recorder operates at $7\frac{1}{2}$ ips, utilizes a $\frac{1}{4}$-inch tape, and has a sampling rate of 48 kHz. Cross-fade error detection allows for razor-blade editing using a channel-mastering machine, much like the X-80. Sony's 24-track PCM-3324 (Fig. 6-42) utilizes $\frac{1}{2}$-inch tape, runs at a speed of 30 ips, and has a 48-kHz sample rate. The 24 digital tracks are located across the width of the tape, with two outside analog tracks and a *control and external data* track located at the center. Error detection for the PCM-3324 permits splice editing, electronic correction, and the execution of a 5.20-millisecond cross-fade.

Fig. 6-42. Sony PCM-3324 digital audio multichannel recorder (*Courtesy Sony Corp.*).

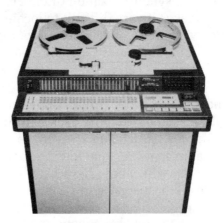

Rotating-head PCM recorders are generally made up of two separate components: a PCM A/D and D/A conversion unit (Fig. 6-43) and a video cassette recorder (VCR). In 1979, the Electronics Industries Association of Japan produced standards designed to ensure the interchangeability of PCM recording equipment that is based on the use of VCRs. In recent years, PCM/VCR combinations have dropped in price, allowing even home audiophiles to produce high-quality digital recordings. Rotating-head PCM recorders have the distinct advantage of being low-cost and are often portable. However, any form of comprehensive editing can only be done electronically since a manual edit would generate too great an error to correct for.

Fig. 6-43. Sony PCM-F1 digital audio processor (*Courtesy Sony Corp.*).

CompuSonics System

Recently, a radical new storage medium and processing method was revealed. CompuSonics of Denver, Colorado, has developed a system of recording, reproducing, and processing digital audio signals which normally would require a large amount of storage space. For storing the tremendous amounts of data involved in recording audio and then compressing this data into a lesser amount of storage space, CompuSonics developed a method called *data reduction*. Data reduction works on the principle that not all information contained in an audio waveform needs be recorded to allow reproduction of the original sound since much of the information contained in an audio sound passage may be inaudible since, among other reasons, it may be very low in level and thereby masked by louder signals. Thus, through data reduction, sounds that are too low in level can be removed from the digital-audio "bit stream" without any change in the perceived sound quality but with a decrease in the amount of data required for storage. The amount of data reduction applied to an audio source is *user selectable* and depends upon the quality of the recorded sound and the program length.

Through data reduction, the recorded data may be recorded directly onto a computer hard or floppy disk (Fig. 6-44). In the case of the DSP-1000 recorder, a 5¼-inch microcomputer-style floppy disk may be able to record up to an hour of program material but, without data reduction, it would hold no more than a few seconds of digital audio information.

Digital Editing

Although many of the stationary-head digital recorders mentioned earlier permit conventional razor-blade edits, most digital PCM-format machines presently on the market require that the splice edit points be made electronically with a digital audio editor. This involves the transfer of the digital audio signal from two playback machines, each containing the separate takes or segments which are to be electronically joined by the editor. The signal is dubbed in final form to a master digital recorder. A digital audio editor (Fig. 6-45) typically includes a control keyboard that is connected to an associated central processor which provides full remote control over the slave playback machines, the electronic editing function, and the master recorder. A time code provides total synchronization and control over selected edit "in" and "out" points. Short-term audio data storage is also provided which stores a short portion of the

Fig. 6-44.
CompuSonics
DSP-2004
system *(Courtesy*
CompuSonics).

audio information from before and after the edit points into the central processor which allows for close examination and checks of the
edit points until a final edit point is located. Once the final decision
is made, the edit can be accomplished easily with all the remote control and edit functions of the tape machine being controlled from the
central processing unit.

**Fig. 6-45.
Mitsubishi XE-1
electronic editor**
*(Courtesy
Mitsubishi
Electric Corp.).*

References

1. *CompuSonics Technical Bulletin,* CompuSonics Corporation.
2. Daniel, Eric D., "Tape Noise in Audio Recording," *Journal of the Audio Engineering Society,* Vol. 20, No. 2, March 1972, pp. 92-99.
3. *M56 Service Manual,* Mincom Division of 3M Company.
4. *Mitsubishi X-800/X-80 Technical Bulletin,* Mitsubishi Electric Corporation.
5. "Preventive Maintenance and Alignment Procedures," *M56 Series 500,* Mincom Division of 3M Company.
6. Schwartz, David M., "Specifications and Explanation of a Computer Audio Console for Digital Mixing and Recording," A paper presented at the 76th Convention of the Audio Engineering Society, October 8-11, New York, 1984.
7. *Sony PCM-3324/3102/F1 Specification Sheet,* Sony Corporation.
8. "Sound Talk," (bulletin) Vol. 1, Nos. 1 & 2, Vol. 2, Nos. 1, 2, & 3, Vol. 3, No. 1, Magnetic Products Division of 3M Company.
9. Tanaka, Kunimara, "The Mitsubishi Digital Audio System," Technical paper.
10. Tremaine, Howard M., "The Audio Cyclopedia," Indianapolis: Howard W. Sams & Co., Inc., 1978.

7 AMPLIFIERS

In the world of audio, the amplifier, as a device, has many applications. It may be designed to amplify, equalize, combine, distribute, or isolate a signal. It may also change the impedance of a signal. At the heart of the amplifier is a regulating device. This regulating device may be either a vacuum-tube or semiconductor-transistor type of device.

Amplification

The *valve*, an original term for the vacuum tube that is still used in England, is a very descriptive word for drawing an analogy of how, in theory, the process of amplification works. In this analogy, assume that we have a high-pressure water pipe (Fig. 7-1), and connected into this pipe is a water valve that is able to control the water pressure with very little effort. By using this valve, we are able to control a large amount of water pressure with a much smaller amount of expended energy.

Fig. 7-1. Water pipe analogy. Current through a transistor is varied in the analogous way that water pressure is controlled by the valve tap of a water pipe.

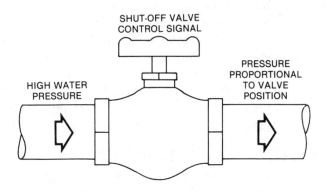

SHUT-OFF VALVE
CONTROL SIGNAL

HIGH WATER
PRESSURE

PRESSURE
PROPORTIONAL
TO VALVE
POSITION

Both the vacuum tube and the transistor work in a very similar manner in that a large dc potential is applied to the input of the device while a smaller control signal is applied to either the grid of the tube or the base of the transistor, causing the device to work much as the water valve does. A large current that is controlled by a small signal input current will appear as a much larger corresponding signal at the output of the device; thus, amplification is accomplished.

The *transistor*, a device commonly used today, is not an inherently linear device. That is, a signal applied to the base will not always produce a corresponding output change. The linear operating region of a transistor lies between the cutoff region and the saturation point of the device (Fig. 7-2). It is in this region that a change in base current produces an equal change in the collector current and voltage. When operating closer to the cutoff and saturation regions, the base current lines are not linear and the output will not be a true facsimile of the input. In order to limit the signal to this operating region, a dc bias signal is applied to the base of the transistor, lifting the signal into the linear region for much the same reason that a high-frequency bias signal is applied to a recording head. Once bias has been applied and sufficient amplifier design characteristics have been met, an amplifier is limited in dynamic range by only two factors: *noise* (which is a result of thermal electron movement within the transistor and its associated circuitry) and *saturation*. Amplifier saturation is the result of the input signal being at such a large level that the dc supply output voltage is not sufficient to produce the required output without severe distortion of the waveform. This produces a process known as *clipping* (Fig. 7-3). For example, if an amplifier has a supply voltage of +24 volts, and is operating with a gain ratio of 30:1, an input signal of ½ volt will produce an output of 15 volts. Should the input be raised to 1 volt, the required output level would be increased to 30 volts. Since the maximum output voltage is limited to 24 volts, wave excursions of greater levels will be chopped off, or *clipped*, at the upper and lower excursions of the waveform, remaining at the maximum level until the signal falls below the maximum level (Fig. 7-3). The result of amplifier clipping is the production of severe odd-order harmonics which, with transistors and many integrated circuit designs, will be immediately audible.

The Operational Amplifier

The *operational amplifier*, or *op amp*, is a stable high-gain high-bandwidth amplifier with a high input impedance and a low output impedance. These qualities allow use of the op amp as a basic building block for a wide variety of audio and video applications; just add

Fig. 7-2. Operating region of a transistor.

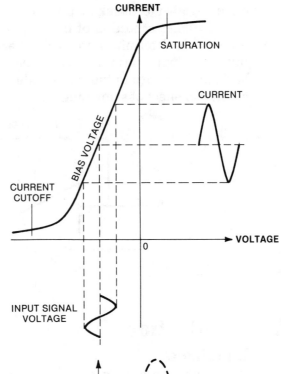

Fig. 7-3. A clipped waveform.

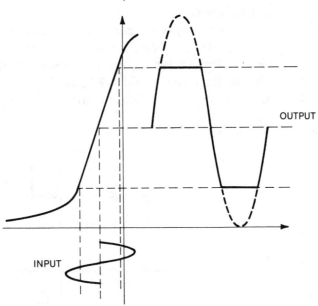

additional components to the basic circuit to fit the required design needs.

Fig. 7-4 shows a typical op amp design used for amplification. In order to reduce the gain of the op amp to more stable workable levels, a negative feedback loop is used. *Negative feedback* is a technique which applies a portion of the output signal through a limiting resistor (which determines the gain) back into the negative or phase inversing input terminal. Thus, part of the output is applied (fed back) out of phase to the input, reducing the overall output. It has the additional effect of stabilizing the amplifier and reducing distortion.

Fig. 7-4. Basic op amp configuration.

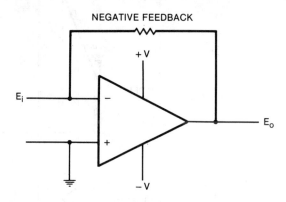

Amplifier Applications

Equalizers

In actuality, an *equalizer* is a frequency-dependent amplifier. In most modern designs, equalization (EQ) is achieved through the use of resistive/capacitive networks located in the negative feedback loop. This is illustrated in Figs. 7-5 and 7-6. By changing the circuit design, any number of equalization curves may be achieved.

Fig. 7-5. Low-frequency equalizer circuit.

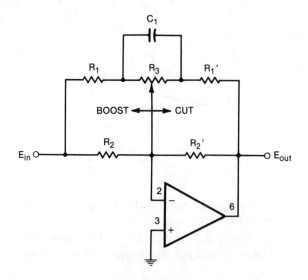

Fig. 7-6. High-frequency equalizer circuit.

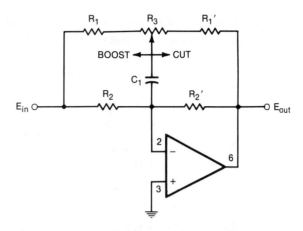

Summing Amplifiers

A *summing amplifier*, also known as an *active combining amplifier*, is designed to combine any number of discrete inputs, while providing a high degree of isolation between these inputs (Fig. 7-7). The summing amplifier is an important component in console design because of the great amount of input signal routing which requires total isolation in order to separate each input from all the other inputs and still maintain signal control flexibility.

Fig. 7-7. Isolation between inputs of a summing amplifier.

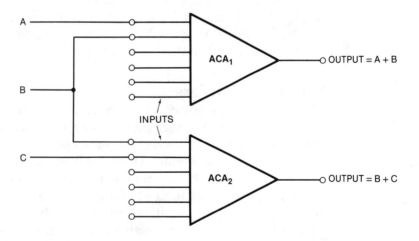

Distribution Amplifiers

The distribution of audio signals to a multitude of devices or signal paths is often necessary. Where increased power (such as in a headphone distribution path) is needed, a *distribution amplifier* is required. Under such circumstances, the amplifier may provide no gain (thus it is termed a unity gain amplifier) but it will amplify the current delivered to one or more loads (Fig. 7-8).

Fig. 7-8.
Distribution
amplifier.

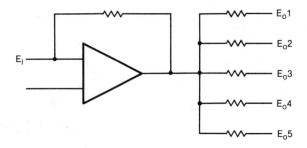

Isolation Amplifiers

An amplifier that is used to isolate signals combined at its input, as in a summing amplifier, may also provide electrical and ground isolation between the output of one device and the input of another. An example of such an amplifier would be an *active direct box* which isolates an electric instrument or amplifier from the console's mike inputs.

Impedance Amplifiers

Another amplifier application is in the changing of the impedance of a signal. The preamp of a condenser microphone provides such an example, where an unusable impedance on the order of one million ohms is reduced to a workable impedance of roughly 200 ohms.

Power Amplifiers

The function of a *power amplifier* (Fig. 7-9) is to boost the current of a signal up to a level where one or more loudspeakers may be driven at the required volume levels. Power amplifiers have their own special set of inherent problems. These include the fact that transistors do not like working at the high temperatures generated by amplifiers during operation at high studio and concert levels. High operating temperatures often result in changes in the response and distortion performance figures. Protective measures, such as fuse and thermal protection, must be taken to ensure reliability at high operating levels and many of the newer amplifier models are able to provide protection for a wide variety of circuit conditions, such as a shorted load, a mismatched load, or an open (no load) circuit. Although these amplifiers will work under various speaker loads, most speaker systems are designed to deliver a source load of 8 ohms.

**Fig. 7-9. Crown
DC-300A Series
II amplifier**
*(Courtesy Crown
International).*

Voltage-Controlled Amplifiers

Up to now, our discussion has centered on amplifiers in which the
output level is directly proportional to the signal level presented at
the input. An exception to this is the *voltage-controlled amplifier* or
VCA (Fig. 7-10). In a VCA, amplifier gain is a function of a dc control
voltage (generally ranging from 0 to 5 volts) that is applied to a control input. Thus, an external voltage may be used to change the level
of an audio signal. Console automation and signal-processing procedures use VCA technology extensively. Voltage-controlled equalizers,
which change the equalization of an EQ amplifier as a function of a
dc control voltage, are also available.

**Fig. 7-10.
Voltage-
controlled
amplifier.**

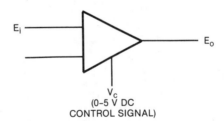

References

1. Crown DC-300A Series II, *Technical Bulletin*, Crown International.
2. Jung, Walter, *Basic Op-Amp Cookbook*, Indianapolis: Howard W. Sams & Co., Inc.
3. Millman, Jacob, and Christos C. Halkias, *Electronic Fundamentals and Applications for Engineers and Scientists*, New York: McGraw-Hill, 1976.
4. Tremaine, Howard M., *The Audio Cyclopedia*, Indianapolis: Howard W. Sams & Co., Inc., 1978.

8 SIGNAL-PROCESSING EQUIPMENT

Equalizers

One of the most important signal-processing devices used in the multitrack studio is the frequency equalizer. This device (Fig. 8-1) gives the engineer control over the harmonic balance or *timbre* of instruments heard by the listener and can be used to compensate somewhat for deficiencies in microphone frequency response or for deficiencies in the sound of an instrument. The frequency equalizer has several uses: to make the sounds from several mikes or several tape tracks blend better than they would otherwise, to match the sound of an overdubbed instrument to the sound of the same instrument when recorded with a different mike or at a different studio, to make an instrument sound completely different from the way it normally would sound for a particular effect, and to increase the separation between instruments by rolling off the leakage frequencies.

Equalization, or *(EQ)*, refers to altering the frequency response of an amplifier such that certain frequencies are more or less pronounced than others. It is specified as plus or minus a certain number of decibels at a certain frequency. Although only one frequency is specified at a time, the frequency set on equalizers used in the studio actually refers to a curve, so signals at 4 kHz and 6 kHz are also boosted somewhat by adding EQ of +4 dB at 5 kHz (Fig. 8-2). The amount of boost or cut at frequencies other than the one named is determined by whether the curve is peaking or shelving, by the Q or bandwidth of the curve, and by the amount of boost or cut at the named frequency.

The equalization available on many older equalizer designs, such as those formerly made by Lang or Pultec, is performed by passive filters, followed by an amplifier used to eliminate the insertion loss of the filters. Fig. 8-3A is a block diagram showing typical signal levels

Fig. 8-1. The Model MEP-130 single-channel, console-mounted unit *(Courtesy International Telecom, Inc.).*

Fig. 8-2. A 5-kHz peak boost curve.

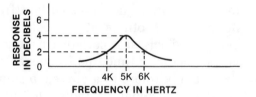

in a passive equalizer set for flat response. The block diagram of Fig. 8-3B shows levels in the equalizer with the filter set for 16-dB boost at 50 Hz. However, most modern design equalization circuits are of the active filter type. By altering the characteristics of the feedback loop in an integrated circuit (Fig. 8-4), it is possible to create any equalization curve without loss of signal level or the noise increase common to passive circuits.

Fig. 8-3. Typical signal levels in a passive equalizer.

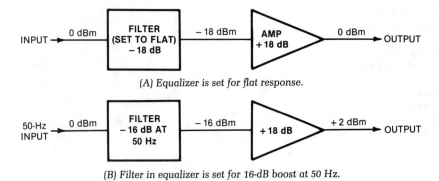

(A) Equalizer is set for flat response.

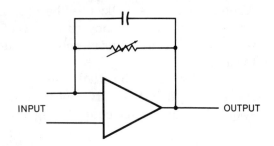

(B) Filter in equalizer is set for 16-dB boost at 50 Hz.

Fig. 8-4. An active equalizer circuit.

Peaking Filters

The most common equalization curve is that of the peaking filter. As its name implies, an actual peak-shaped bell curve is created which may be either boost or cut at a selected center frequency. Fig. 8-5 shows the curves for a peak equalizer set to boost or cut at 1000 Hz. The Q of a peaking equalizer refers to the width of the bell-shaped curve (Fig. 8-6). A curve with a high Q will have a narrow bandwidth, with few frequencies outside the selected bandwidth being affected, whereas a low-Q curve is very broadband, affecting many frequencies.

Fig. 8-5. A bell curve.

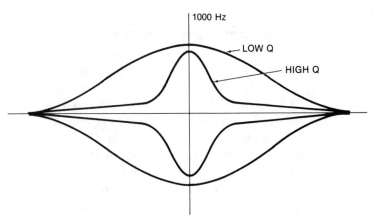

Fig. 8-6. The number of hertz between the points which are 3-dB down from the center frequency is the bandwidth (BW) of a peaking filter.

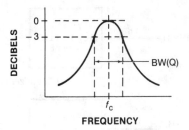

Shelving Filters

Another type of equalizer is the shelving filter. *Shelving* refers to a rise or drop in frequency response, at a selected frequency, which tapers off to a preset level and continues, at this level, to the end of the audio spectrum.

Shelving can be inserted at either the high or low end of the audio spectrum and is the curve most commonly displayed by the home system bass and treble controls (Fig. 8-7).

Fig. 8-7. High/ low boost/cut curves for a shelving equalizer.

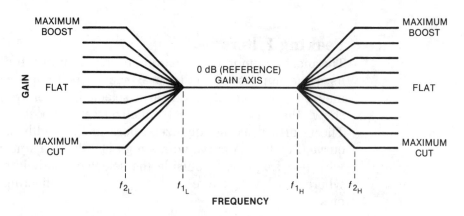

High- and Low-Pass Filters

Other types of equalizers are the *high-pass* and *low-pass* filters. As their names imply, certain frequencies are passed at full level, while others are attenuated. Frequencies which are attenuated by less than 3 dB are said to be inside the *passband*, while those attenuated by more than 3 dB are in the *stop band*. The frequency at which the signal is attenuated by exactly 3 dB is called the *turnover* or *cutoff frequency* and is used to name the filter. Ideally, attenuation would become infinite immediately outside the passband but, in practice, this is not attainable. In the simplest case, attenuation increases at a rate of 6 dB per octave. This rate is called the slope of the filter.

Other slopes in common use are 12 and 18 dB per octave. For example, Fig. 8-8 shows a 700-Hz high-pass filter response curve with a slope of 6 dB per octave, while Fig. 8-9 shows a 700-Hz low-pass filter response curve with a slope of 12 dB per octave.

Fig. 8-8. A 700-Hz high-pass filter with a slope of 6 dB per octave.

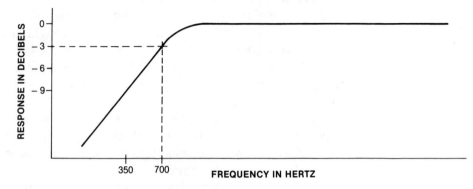

Fig. 8-9. A 700-Hz low-pass filter with a slope of 12 dB per octave.

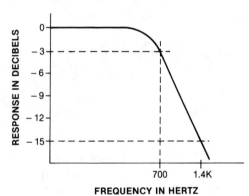

High and low-pass filters differ from shelving EQ in that their attenuation does not level off outside the passband, rather it continues to increase. A high-pass filter in combination with a low-pass filter can be used to create a bandpass filter with the bandwidth being controlled by their turnover frequencies and the Q controlled by their slopes (Fig. 8-10).

Fig. 8-10. A bandpass filter created by combining high- and low-pass filters with different cutoff frequencies.

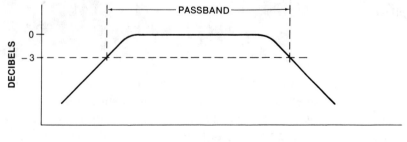

Equalizer Types

There are four basic types of equalizers in use today: the selectable frequency equalizer, the parametric equalizer, the graphic equalizer, and the notch filter.

The *selectable frequency equalizer*, as its name implies, has a predetermined set number of frequencies from which to choose. The selected frequency will usually allow a boost or cut with a predetermined Q bandwidth and, typically, will have an independent low/mid/high range. This form of equalization is most often found on older console designs and on newer low-cost production consoles.

The *parametric equalizer* (Fig. 8-11) differs from the selectable equalizer in that it has the advantage of a continuously variable sweep control over the selection of frequency, boost, and cut and, often, control over the bandwidth Q. The parametric equalizer has, over the last five years, become the standard design incorporated into the input strips of modern recording consoles.

Fig. 8-11. The Symetrix SE-400 stereo parametric equalizer (*Courtesy Symetrix*).

A third type of equalizer is the *graphic equalizer* (Fig. 8-12). This equalizer is generally made up of 12 to 36 slider controls, oriented vertically, such that each slider control will peak boost or cut one specific frequency band. With all the sliders oriented in this manner, a visual representation of the equalization curve is clearly visible. Graphic equalizers find a particular use where there is a need for the acoustical fine tuning of a room, such as a control room or an auditorium.

Fig. 8-12. The Rane graphic equalizer (*Courtesy Rane Corp.*).

In addition to its use in modifying sound, an equalizer can be used to remove hum and other undesirable discrete-frequency noises. A *notch filter* is used for this purpose (Fig. 8-13). The filter can be tuned to attenuate a particular frequency and has a very narrow bandwidth so it has little effect on the rest of the program (Fig. 8-14).

Notch filters are used more in film-location sound than in studio recording because of the problems encountered in location work which are usually not present in a well-designed studio.

Fig. 8-13. The Universal Audio Model 565 Little Dipper variable high- and low-pass, peaking, and notch filter *(Courtesy United Recording Electronics Industries).*

Fig. 8-14. Frequency response of a notch filter.

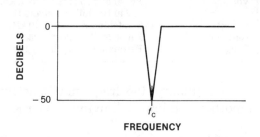

Equalization must be done by ear, but it is helpful to have an idea of what frequencies will give the desired effect. Useful EQ frequencies for some common instruments are provided in Table 8-1, and the frequency ranges of the instruments are illustrated in Fig. 8-15.

Fig. 8-15. Frequency ranges of instruments.

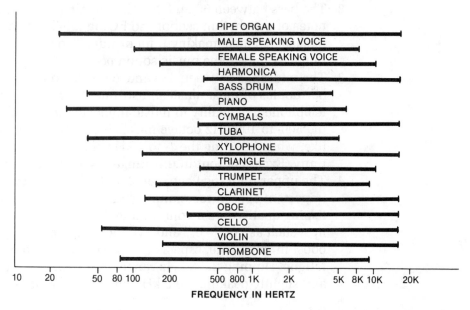

**Table 8-1.
Useful EQ
Frequencies for
Common
Instruments**

Instrument	Frequency Ranges of Interest
Bass guitar	Attack or pluck is increased at 700 or 1000 Hz, bottom added at 60 or 80 Hz, string noise at 2.5 kHz.
Bass drum	Slap at 2.5 kHz, bottom at 60 or 80 Hz.
Snare	Fatness at 240 Hz, crispness at 5 kHz.
Hi hat and cymbals	Shimmer at 7.5 kHz to 10 kHz. Clank or gong sound at 200 Hz.
Tom toms	Attack at 5 kHz, fullness at 240 Hz.
Floor toms	Attack at 5 kHz, fullness at 80 or 120 Hz.
Electric guitar	Full at 240 Hz, bite at 2.5 kHz.
Acoustic guitar	Body at 240 Hz, clarity at 2.5 kHz, 3.75 kHz, and 5 kHz, with the sound thinning out as frequency rises. Bass strings at 80 or 120 Hz.
Organ	Bass 80 to 120 Hz, presence 2.5 kHz, body 240 Hz.
Piano	Bass 80 to 120 Hz, presence 2.5 kHz to 5 kHz, thinning as frequency is raised. Honky-tonk sound at 2.5 kHz as bandwidth is sharpened. Resonance and echo at 25 to 50 Hz.
Horns	Fullness at 120 to 240 Hz, shrill at 7.5 or 5 kHz.
Voice	Presence at 5 kHz, sibilance at 7.5 kHz to 10 kHz, boominess at 200 to 240 Hz, fullness at 120 Hz.
Strings	Scratchiness at 7.5 to 10 kHz, fullness at 240 Hz.
Harmonica	Fat at 240 Hz, electric at 2.5 kHz, acoustic at 5 kHz.
Conga	Resonant ring at 200 to 240 Hz. Presence and slap at 5 kHz.

In summary, the frequency spectrum can be divided up into six important sections, following de Gar Kulka's description:[4]

1. The very low bass between 16 and 60 Hz which encompasses sounds that are often felt more than heard, such as thunder in the distance. These frequencies give the music a sense of power even if they occur infrequently. Too much emphasis on this range makes the music sound muddy.

2. The bass between 60 and 250 Hz contains the fundamental notes of the rhythm section, so EQing this range can change the musical balance, making it *fat* or *thin*. Too much boost in this range can make the music sound *boomy*.

3. The midrange between 250 and 2000 Hz contains the low-order harmonics of most musical instruments and can introduce a telephone-like quality to music if boosted too much. Boosting the 500- to 1000-Hz octave makes the instruments sound horn-like, while boosting the 1- to 2-kHz octave makes them sound tinny. Excess output in this range can cause listening fatigue.

4. The upper midrange between 2 and 4 kHz can mask the important speech recognition sounds if boosted, introducing a lisping quality into a voice and making sounds formed with the lips, such as "m," "b," and "v" indistinguishable. Too much boost in this range, especially at 3 kHz, can also cause listening fatigue. Dipping the 3-kHz range on instrumental backgrounds and slightly peaking the 3 kHz on vocals can make the vocals

audible without having to decrease the instrumental level in mixes where the voice would otherwise seem buried.

5. The presence range between 4 and 6 kHz is responsible for the clarity and definition of voices and instruments. Boosting this range can make the music seem closer to the listener. Adding 6 dB of boost at 5 kHz makes a mix sound as if the overall level has been increased by 3 dB. As a result of this effect, many record companies and mastering engineers make a practice of adding a few dB of boost at 5 kHz to make their product sound louder. Reducing the 5-kHz content of a mix makes the sound more distant and transparent.

6. The 6- to 16-kHz range controls the brilliance and clarity of sounds. Too much emphasis in this range, however, can produce sibilance on vocals.

The best way to learn to use an equalizer is to set the amount of boost to a maximum and change the boost frequency until the desired range of the instrument to be EQed is found. The amount of boost can then be decreased until the desired effect is obtained. Attenuation of a frequency range can be achieved in a similar manner. Drawing a curve of the effect introduced by the equalizer when the desired sound is attained can often aid in future equalizing and in visualizing what the controls actually do.

If boosting one range of an instrument creates the need to boost the other ranges, and they are then boosted, the effect achieved is simply that the overall level has been raised. This is more easily done with the input fader. If the increased fader level does not make the sound satisfactory, it may be that one range of frequencies is too dominant and requires attenuation. Just because the ear can hear from 20 Hz to 20 kHz does not mean that these frequencies should exist on every record. There are many times that the presence of these frequencies may detract rather than add to the desired effect.

As far as recording with EQ goes, there are different opinions. If an engineer other than one who records the multitrack tape is to mix it, he may have a very different idea of how the instruments should sound and may have to work very hard to counteract the EQ used by the original engineer. If everything is recorded flat, however, the producer and artists will have to strain, while they are trying to pass judgement on a performance, to imagine how the instruments will sound later. It is also important to know how the instruments will sound with EQ during overdubbing so that the producer and artists can decide when a song has been "sweetened" enough. When several mikes are to be combined onto one channel of the tape, they can be EQed individually only before recording, so that recording flat, as a rule, will prevent later optimization during the mixdown of sounds picked up by each mike. In addition, while recording with EQ does not change the perceived noise level, playing back with EQ does. EQ used on playback is also added to the residual tape noise of the track. So boosting highs, for example, during playback, would make the tape hiss of that channel more audible than if the highs were boosted before recording. If the same engineer is to record and mixdown the tape, then recording with EQ is usually not a problem. In any event, unless a special effect is desired, EQ should be used moderately, and microphone selection should be used to obtain a good instrument sound. If an instrument is poorly recorded in an initial recording session, it can rarely be corrected later during the mixing.

Meters, Compressors, Limiters, and Expanders

Meters

Since amplifiers and magnetic tape are limited in the range of signal levels they can pass without distortion, audio engineers need a means of determining whether the signals they are working with will be stored or transmitted without distortion.

The most convenient method of indicating this is through the use of a visual device such as a meter. If preventing distortion on the tape were the only concern, peak-indicating meters could be used to display the maximum amplitude fluctuations of a waveform. The human perception of loudness, however, does not have much relationship to the peak level of signals. The meter may read higher at a certain point in the program, but the program may not sound any louder (Fig. 8-16). If the meter is used to set or maintain a certain volume level, peak indication is of no use.

Fig. 8-16. A peak meter reads higher at point A than at point B, even though the loudness is the same.

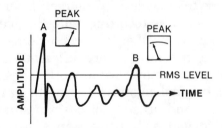

Since the perception of the ear to loudness is proportional to the *rms value* of a signal, a meter was designed that could read this level so that volume and meter indication would coincide (Fig. 8-17). The scale chosen for the meter was calibrated in volume units; hence the name VU meter (Fig. 8-18). Zero VU is considered the standard operating level. Although VU meters do the job of indicating volume level, they ignore the short-term peaks which can overload tape. These peaks can be from 8 to 14 dB higher than the rms value indicated. This means that the electronics must be designed so that unacceptable distortion does not occur until at least 14 dB above 0 VU. Typical VU meter specifications are given in Chart 8-1.

Fig. 8-17. A VU meter reads the rms level and ignores the peaks that do not contribute to loudness.

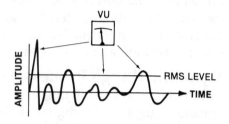

Fig. 8-18. A VU meter. The upper scale is calibrated in volume units for use in recording. The lower scale is percentage modulation for use in broadcasting.

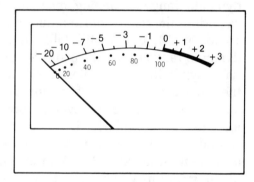

Chart 8-1. VU Meter Specifications

Sensitivity:	Reads 0 VU when connected across a +4-dBm signal (1.228 volts in a 600-ohm circuit).
Frequency Response:	±0.2 dB from 35 Hz to 10 kHz. ±0.5 dB from 25 Hz to 16 kHz.
Overload Capability:	Can withstand ten times 0-VU level (+24 dBm) for 0.5 second, and five times 0 VU (approximately +18 dBm) continuously.

Volume units indicated on the meter are equal to dB for sine waves, but for other waves or complex signals, the VU meter actually reads between the rms and peak values of the signal. For these waves, volume units are larger than dB.

The difference between the maximum level which can be handled without excess distortion and the average operating level of the system is called *headroom*. Some studio-quality amplifiers are capable of outputs as high as 26 dB above 0 VU and thus have 26 dB of headroom. Magnetic tape, however, has limited headroom, for its dynamic range is such that providing the headroom necessary to prevent distortion of the peaks would make tape noise too audible during the rest of the program. The 3% distortion level for magnetic tape recorded on a tape machine that does not use a linearizer is only 6 dB above 0 VU, while console amplifiers have distortion of less than 0.4% at this level. The proper record level for most program material is 0 VU although higher levels are possible providing that peaks, which would cause distortion, are not present.

Meters are now available that use *light-emitting diodes (LEDs)* or *liquid-crystal displays* to provide level indication through the illumination of lights corresponding to different levels, rather than through the use of a meter pointer. These units follow peaks better than any meter can, giving virtually instantaneous display of the signal level. Often these indicators are switchable to read peak or rms levels.

Dynamic Range Processors

The dynamic range of music is on the order of 120 dB, while the dynamic range of the digital medium is on the order of 90 dB. The dynamic range of analog magnetic tape is on the order of 60 dB, excluding the use of noise-reduction systems which can add another 15 to 30 dB, but it still falls short of the 120 dB of music. The dynamic range of an LP record is about 70 dB. However, often with a wide dynamic range, unless used in a noise-free environment, either the quiet passages are lost in the ambient noise of the listening area (35- to 45-dB spl for the average home), or the loud passages are too loud to bear. Similarly, if a program of wide dynamic range were reproduced through a medium with a narrow dynamic range, such as an a-m radio (20 to 30 dB) or an fm radio (40 to 50 dB), a great deal of information would be lost in background noise. To prevent these problems, the dynamic range is reduced to a level appropriate both for the medium through which it is to be reproduced and for comfortable listening in the average home. This reduction is accomplished by a combination of the engineer *riding the faders* and the use of a device called a *compressor*.

Compressors

A compressor is, in effect, an automatic fader. When the input signal exceeds a predetermined level (Fig. 8-19A), called the *threshold*, compressor gain is reduced and the signal is attenuated (Fig. 8-19B). The number of dB increase of input signal needed to cause a 1-dB increase in the output signal of the compressor is called the *compression ratio* or the *slope* of the compression curve. Thus, for a ratio of 4:1, an 8-dB increase of input produces a 2-dB increase in the output. Since the signals generated in music vary in loudness and, therefore, may be above the threshold at one instant and below it the next, the speed with which the gain is to be reduced, after the signal exceeds the threshold, and then restored, after the input signal falls below the threshold, must be determined. These speeds are determined by the *attack* and *release* times, respectively.

Fig. 8-19. The compressor reduces the level of the portion of the signal which exceeds the threshold.

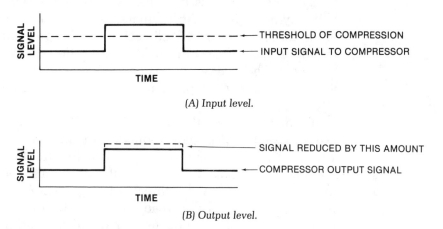

(A) Input level.

(B) Output level.

As stated earlier, the perception of the ear to the loudness of a signal is proportional to rms value, so large short-duration peaks do not noticeably increase the loudness of a signal. What is desired is to allow the signal volume to rise and fall but to lesser extents than the volume would if it were not controlled. If the peaks of the waveform were permitted to trigger *gain reduction*, the volume would actually decrease, rather than increase by a smaller amount. This would change the dynamics of the program noticeably and in an unacceptable way. To avoid the triggering of compression by these peaks, the attack time is set so that the waveform must exceed the threshold level for a time long enough to constitute an increase in the average level, and gain reduction will not decrease the overall volume. The attack time is defined as the time it takes for the gain to decrease by a certain amount, usually to 63% of its final value.

If the release time was set too short for the program material, i.e., if full gain was restored each time the signal fell below the threshold, *thumps, pumping,* and *breathing* would be heard due to the rapid rise of background noise as the gain was increased. Also, if a rapid succession of peaks were fed into the device, the program gain would be restored after each one, and the level of the program would be heard to rise after each peak. Since the level-sensing mechanism is sensitive to both positive and negative waveform excursions, extremely short release times could cause gain reduction twice each cycle, introducing harmonic distortion into the signal.

To eliminate these effects, longer release times are used so that repeated waveform excursions past the threshold cause gain reduction only once. The gain remains reduced through all these excursions and returns to normal gradually. This makes the increase in background noise less obvious, as well as making any gain changes that may be required shortly thereafter less drastic. If the release time is too long, however, a loud section of the program may cause gain reduction that persists through a soft section, making the soft section inaudible. The release time is defined as the time needed for the gain to return to a certain percentage of its no-gain reduction value (usually 63%).

Compressors usually have a built-in VU meter to allow monitoring of the amount of gain reduction taking place. The meter usually sits at 0 VU when the input signal is below the threshold and falls to the left to indicate the number of dB of gain reduction when the input signal exceeds the threshold (Figs. 8-20A and B). Some compressors use meters that read gain reduction directly, resting at the left side of the scale until the signal exceeds the threshold and then moving upscale to show gain reduction in dB (Figs. 8-21A and B). On some compressors, the meter is switchable to read either gain reduction or output level.

Fig. 8-20. The gain-reduction meter on a compressor.

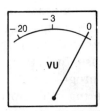

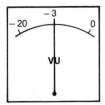

(A) The input signal to the compressor is below the threshold and compressor gain is normal.

(B) The input signal exceeds the threshold causing 3 dB of gain reduction.
The meter indicates that the gain is 3 dB lower than normal.

Fig. 8-21. The gain-reduction meters on some compressors are designed to read gain reduction directly.

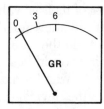

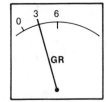

(A) No gain reduction.

(B) Gain reduction of 3 dB.

Limiters

If the compression ratio is made large enough, the compressor becomes a *limiter* (Fig. 8-22). A limiter is used to prevent signal peaks from exceeding a certain level in order to prevent overloading amplifiers, tapes, or discs (Fig. 8-23). An extreme case of a limiter is a *clipper* which chops off the top of any waveform exceeding the threshold level. A clipper could be said to have an infinite compression ratio. Most limiters have ratios of 10:1 or 20:1, although they are available with ratios up to 100:1. Since such a large increase in the input signal is needed to produce an increase in the output of the limiter, the likelihood of overloading the equipment following the limiter is greatly reduced.

Fig. 8-22. A stereo compressor-limiter-expander *(Courtesy Audio + Design).*

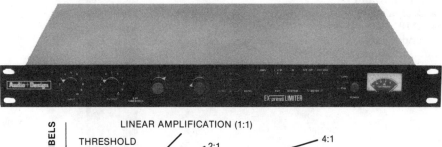

Fig. 8-23. The output of a compressor is linear below the threshold point and follows the slope of the selected compression or limiting curve above the appropriate threshold.

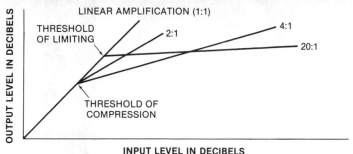

Limiting is most often used in recording to prevent short-term peaks from reaching their full amplitude. Short-term peaks add little information to the program in proportion to the distortion they would cause (if they saturated the tape) or the noise they would allow to enter the system (if the signal were recorded at a level low enough so that the peaks would not distort). Extremely short attack and release times are used so that the ear cannot hear the gain being reduced and bought back up. Limiting is used to remove only occasional peaks, for gain reduction on many successive peaks would be noticeable. If the program contains many peaks, the threshold should be raised and the gain reduced manually so that only occasional extreme peaks are limited.

Expanders

Expansion is the process of decreasing the gain of a signal as its level falls and/or increasing the gain as the level rises. Thus, when the signal level is low (below the expansion threshold), gain is low and program loudness is reduced. On certain expanders, when the signal level increases above the threshold, the gain is increased. Expanders (Fig. 8-24) increase the dynamic range of a program by making loud signals louder and soft signals softer. They also can be used as noise-reduction devices by adjusting them such that the noise to be removed is below the threshold level, while the desired signal is above the threshold.

Fig. 8-24. Audio + Design's S30 SCAMP expander-gate *(Courtesy Audio + Design).*

Noise Gates

Another type of expander is the *noise gate* or, simply, *gate*. This device will allow a signal which is above the threshold to be passed through to the output, while effectively turning off when the signal falls below the threshold. Thus, the desired signal is allowed to pass while background noise is not. This device is also very effective in cutting down leakage on drums and other multi-miked instruments. Often a keyed input is included, which permits another signal source, such as an oscillator or another instrument, to allow the actual program gating.

Summary

The selection of proper attack and release times and degrees of compression, expansion, and limiting depends on the program material. In multitrack recording, dynamic range modifications usually deal with only single instruments or groups of instruments. In radio, television, and disc cutting, entire songs are compressed and the problem of using the proper parameters is more critical.

Limiting is usually only used for recording speech or instruments with *transients* (momentary high peak levels) so that the signal can be recorded at a high level without overloading the tape.

Compression is used for several reasons:

1. It minimizes the change in volume which occurs when an instrumentalist or vocalist momentarily changes his distance from a mike, or has too great a dynamic range for the music.
2. It can balance out the different volume ranges of an instrument. For example, some bass guitar strings are usually louder than the others on the same guitar, and the use of compression produces a smoother bass line by matching the volumes of the different notes. As another example, some instruments, such as horns, are louder in some registers than in others due to the amount of effort required to produce the notes. Compression equalizes the volume levels of the different registers.
3. Compression enables a signal to be made significantly louder in the mix, while increasing the overall signal-level reading on the meter only slightly. This is accomplished by increasing the ratio of average-to-peak levels.
4. Compression can be used to reduce sibilance in a voice by inserting a filter in the compression circuit which causes it to trigger compression when an excess of high-frequency signal is present. A compressor used in this manner is often called a *de-esser*.

Producers often strive to cut their records as *hot* as possible; i.e., they want the recorded levels to be as far above the normal operating level as possible without blatant distortion. They talk about *competitive levels*, for they feel that the louder records will stand out and sound better than the soft ones when a stack is being played on a record changer or over "top-40" radio. In fact, a record that is one or two dB louder than another will sound better due to the Fletcher-Munsen effect which makes the louder record appear to have more bass and more highs. To achieve these hot levels without distortion, compressors and limiters are often used in disc cutting to remove peaks and raise the average level of the program so that the disc would be louder than it otherwise would be. A combinational unit is shown in Fig. 8-25.

Fig. 8-25. A peak-rms compressor/ limiter *(Courtesy Symetrix).*

Compressing a mono mix is done in the same manner as compressing a single instrument, except that the adjustment of the threshold, attack, release, and ratio controls is more critical in preventing *pumping* due to instruments that are prominent in the mix. Compressing a stereo mix gives rise to an additional problem. If two individual compressors are used, a peak in one channel will reduce the gain on that channel, causing sounds centered between the two speakers to jump towards the channel not undergoing compression. To avoid this *center shifting*, most compressors have a provision for connecting them in stereo with a second compressor of the same make and model. This interconnection procedure mixes the outputs of the signal-level sensing circuits of the two units, so that a signal which causes gain reduction in one channel will also cause equal gain reduction in the other channel, preventing the center information in the mix from shifting.

Several signal processing manufacturers have adopted a design which allows compact signal-processing modules to be fitted into a single rack-mount frame, permitting a flexibility that is tailored to the user's present and future needs. One such system is the SCAMP, a standardized compatible audio modular package (Fig. 8-26) from Audio + Design, Calrec Inc. The SCAMP system was conceived to provide a wide range of signal-processing equipment in a space-saving, high-density format. Up to seventeen 1-inch modules can be accommodated in the 19-inch rack, with several modules having a dual channel capability. The following modules are available under the SCAMP system: an expander/gate, a dual noise gate, a dynamic-filter noise gate, a de-esser, a compressor/limiter, a sweep EQ, a parametric EQ, an octave EQ, a crossover/4-band processor, a mike pre-amp, a distribution amp, a delay/time shape, and panning effects. Two examples of the compressors and limiters in common use are the UREI Model 1176 LN peak limiter and the UREI Model LA-4 compressor/limiter.

Fig. 8-26. SCAMP (Standardized Compatible Audio Modular Package), a system with a full complement of modules *(Courtesy Audio + Design, Calrec Inc.).*

The UREI Model 1176 LN peak limiter (Fig. 8-27) has compression ratios of 4:1 and 8:1, as well as ratios of 12:1 and 20:1. It has adjustable attack and release times and an adjustable output level after the gain-reduction section. The amount of gain reduction is selected by setting the meter selector switch to GR and advancing the input level control until the desired gain reduction is either heard or indicated on the meter. The output level control is then adjusted so that the desired level is presented to the input section of the console. The output level can be monitored by depressing either the +4 or +8 meter-selection switch. The +4 switch calibrates the VU meter so that 0 VU equals +4 dBm, while with the +8 switch, 0 VU equals +8 dBm. It is good practice for the output level of all compressors to be set so that their output does not exceed 0 VU on the highest meter-calibration scale in order to prevent overloading their output amplifiers.

Since very fast release times (from 50 milliseconds to 1.1 seconds) are available with the Model 1176 LN, precautions must be taken when compressing or limiting low-frequency signals. If the release time is set too short, the compressor action can follow the individual cycles of the waveform, causing harmonic distortion. As a guideline, when compressing bass guitar or other instruments with a high proportion of low frequencies, the release-time control should not be set for too short a release time. On other material, short release times

Fig. 8-27. The UREI Model 1176 LN peak limiter (*Courtesy JBL/UREI*).

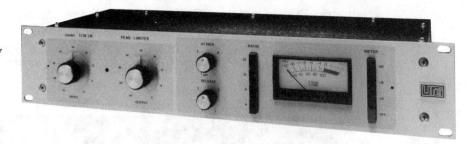

will not cause harmonic distortion, but may cause the compressor action to become noticeable due to obvious level changes.

The UREI Model LA-4 compressor/limiter (Fig. 8-28) is fitted with a meter and four front-panel controls, consisting of meter switching to allow output or gain-reduction monitoring, a peak-reduction or gain-reduction control (which determines how much gain reduction will take place), an overall output gain control (which adjusts the output level of the device), and a power switch.

Fig. 8-28. The UREI Model LA-4 compressor/ limiter (*Courtesy JBL/UREI*).

The Model LA-4 has a rear-panel gain switch which adjusts the gain to either 20 or 40 dB. A compress/limit switch allows the unit to be used as either a compressor or as a 20:1 limiter. Attack time is between 1 to 10 milliseconds and is determined by the program material. The release time varies from 100 milliseconds to 1 second, depending on the duration of the peak causing the gain reduction. There is also a rear-panel control that can increase the gain reduction at high frequencies relative to that below 1 kHz, as well as adjustments to equalize the gain reduction of two units interconnected for stereo. Because of the small size of the unit, two LA-4 units may be mounted side by side in a standard 19-inch rack.

An example of an expander is the Kepex expander (Fig. 8-29A). It acts somewhat like a compressor in reverse. With no input signal, gain reduction is at the maximum preset amount (variable from 0 to 60 dB). When the input-signal level exceeds the threshold, gain is increased to unity. The Kepex expander has a fixed attack time of less than 20 microseconds and a variable release time from 50 milliseconds to 6 seconds. The threshold level and amount of expansion are both variable and the expansion ratio varies from 1:2 to 1:4, depending on the setting of the expansion control. The Kepex unit can tighten up signals by removing echos and rings from them by reducing the gain after the major portion of the signal is over (Fig. 8-29B). It has the additional capacity to *key* or initiate the expansion of one signal on the basis of a second signal for special effects. Peak values of gain reduction are indicated by lights behind the front-panel scale. Threshold-type expanders can also be used as *noise gates* because they can be set to permit a signal to pass and then shut down the channel when no signal is present to keep out noise.

Gain reduction can be achieved in several ways. The Model LA-4 uses an *electroluminescent attenuator* which is a phosphorescent panel that glows in proportion to the input signal fed to it. This glow illuminates two photocells which are part of a volume control circuit (Fig. 8-30). The value of the attenuator changes with the illumination received by the photocells and thus turns the volume of the signal up or down. The Kepex and 1176 LN use a *field-effect transistor (FET)* as a *voltage-variable resistor*. The FET changes resistance in proportion to a voltage (obtained from the input signal) applied to it and this resistance change is used to control the signal in the same manner as the photocells.

Another method, which is gaining acceptance, uses a different method of gain control called *pulse-duration modulation (PDM)*. This method feeds the signal through an electronic switch that is opened and closed at the rate of 200 kHz (Fig. 8-31). By varying the amount of time that the switch is left open or closed during each cycle, energy is removed from the signal without affecting the program, and the result is an output signal of lower level than the input signal.

Fig. 8-29. Kepex II expander and curves.

(A) Kepex II expander (Courtesy Valley People, Inc.).

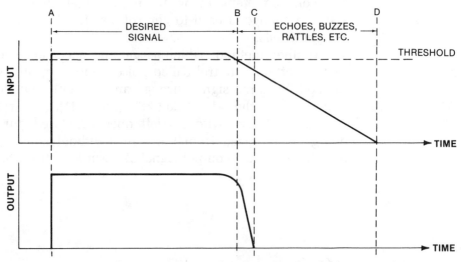

(B) Using the Kepex to remove echoes, buzzes, and rattles from a signal
(Courtesy Allison Research, Inc.).

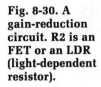

Fig. 8-30. A gain-reduction circuit. R2 is an FET or an LDR (light-dependent resistor).

Fig. 8-31. Operation of the PDM system.

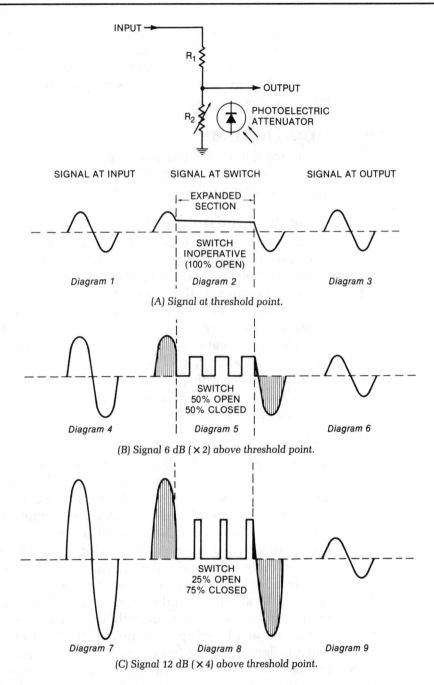

(A) Signal at threshold point.

(B) Signal 6 dB (× 2) above threshold point.

(C) Signal 12 dB (× 4) above threshold point.

Special Effects

In addition to EQ, compression, and expansion, there are many other devices or processes which can be used to make the original signal sound like something other than what it really is, which can add a great deal of interest to a song. A signal processed in such a manner is called an *effect*. The five most common effects are created through the use of reverberation, time delay, phasing, harmonizing, and digital processing.

Reverberation Devices

There are five major sources of reverberation: (1) acoustic environment, (2) acoustic echo chambers, (3) spring reverb devices, (4) echo plates, and (5) digital reverberation.

Acoustic Environment

Acoustic environment makes use of the natural reverberation present in a concert hall or a "live" room. Under certain circumstances, the actual environment will give the most exciting and natural sound. The best example of this would be the sound of a concert hall. This environment was designed to give a natural reverberation to specific forms of music. When recording, the environment could figure very strongly in the overall sound placed on tape. Other environments can be just as important, such as the natural sound of a live recording studio in the recording of drums, or guitars used for overdubs. One studio has actually built a totally stainless-steel-lined room for modern music production.

Acoustic Echo Chambers

Another form of reverb device is the acoustic echo chamber (Fig. 8-32). This consists of a room with highly reflective surfaces, in which a speaker and microphone are placed. The speaker is fed the signal to be reverberated and the mike picks up a combination of direct sound from the speaker and reflections off the walls, ceiling, and floor. By using a directional mike and by placing the mike and speaker back to back pointing in opposite directions, the pickup of direct sound can be minimized. Movable partitions can be set up in these rooms to vary the decay time. Good acoustic echo chambers have a very natural quality to them when properly designed. Their disadvantages are that they use a lot of space (typically 18 by 15 by 12 feet), require as much isolation from external sounds as a studio, and need high-quality speakers and mikes. This makes the price of an acoustic chamber very high. Smaller rooms can be used but, generally, the bass response suffers. Size and cost have generally made this device an unviable option, especially with the calibre and quality of the electronic reverb devices that are available today.

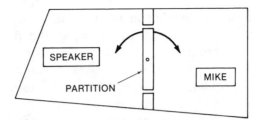

There are, however, certain chambers that have such a natural quality to them, such as the Capital Studios chambers in Los Angeles, that other studios will rent usage time in them and will transmit and receive their signals via equalized telephone lines.

Spring Reverb Units

A third type of reverberation device is the spring reverb unit (Fig. 8-33). This device consists of a spring or set of springs, a driver which is used to set the spring assembly in motion (creating waveforms which travel back and forth along the spring's length), and a pickup which receives and amplifies the reverberated signal. This will often be followed by an equalization circuit to even and smooth out the signal. The propagation velocity of the wave in the springs is much higher than the speed of sound in air and varies with the thickness of the spring wire, the number of turns to an inch, and the length of the spring. The spring length is usually about one foot per set.

Fig. 8-33. A spring reverb unit.

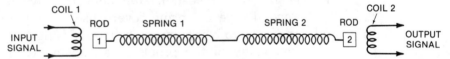

In practice, several sets of springs are used, and each set consists of at least two springs connected in series. The series springs are chosen to have different wire sizes and different numbers of turns to the inch. When the wave reaches the far side of the springs, it causes a set of magnetic rods to rotate and induce a current flow in a coil which converts the rotating mechanical motion into electrical current, which is then amplified and becomes the reverb signal (Fig. 8-33). The signal does not just pass through the springs just once, however. At each junction (rod to spring, spring to spring, and spring to rod), part of the signal is reflected back through the spring it has just flowed through, so that it takes longer for certain parts of the input signal energy to reach the output of the springs. The springs dissipate energy in the form of heat, so that the waves which have followed longer paths are lower in level than the other waves by the time they reach the output. Since the springs are different, they create different delay times, so the net effect is that of many reflections coming shortly after the input signal, closely spaced but decreasing both in time between reflections and in level as time passes (Fig. 8-34). The decay time of this type of spring unit can be decreased by bringing some acoustic absorbing material, such as fiberglass, near the springs to absorb the energy of the air molecules around the springs and damp their rotating motion.

Fig. 8-34. The signal path in a spring reverb unit. A reflection occurs at each boundary: spring to spring and spring to rod.

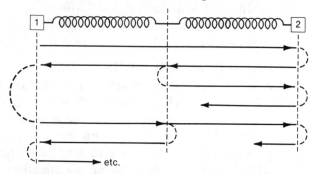

Many spring reverb units suffer from the problem of sounding tinny, unnatural, and, in general, very reminiscent of springs. Recent designs have attempted to eliminate many of the problems through carefully controlled spring lengths and equalization to achieve smooth frequency response. These units have the advantage of small size and low cost, making them attractive for small studios.

A second type of spring reverb uses springs hanging vertically and connected to a moving coil driver. Using this design, AKG has overcome many of the problems associated with spring-type echo chambers in their Models BX25E (Fig. 8-35), BX20E (Fig. 8-36), and BX10 through the use of specially designed spring and damping elements. The units consist of two identical reverberation channels which can be used either independently as two mono chambers (separation between them is better than 60 dB) or used with their inputs in parallel to create stereo reverb. The decay time on the BX20E is variable from 2 to 4.5 seconds and is controlled by electronic damping which can be adjusted remotely and separately by two dc voltages. Since the decay time is varied electronically, it can be changed during a program without causing additional noise. Fig. 8-37 shows the remote control unit for the BX25 reverb unit. The AKG BX25E, BX20E, and BX10 reverberation units are insensitive to physical shock, enabling them to be moved for location recording without taking excessive precautions to prevent damage to the springs. They are well isolated from ambient noise and, if desired, may be left in the control room; acoustical feedback will not occur until the sound level of a unit's output through the monitor speakers is greater than 100-dB spl near the unit.

Fig. 8-35. AKG BX25E spring reverberation unit (*Courtesy AKG Acoustics, Inc.*).

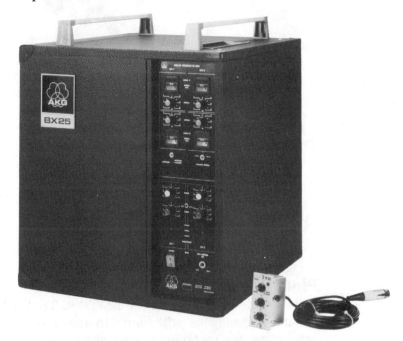

Fig. 8-36. Interior view of the AKG BX20E reverberation unit (*Courtesy North American Philips Corp.*).

**Fig. 8-37. AKG
BX25E remote
control unit**
*(Courtesy
Acoustics, Inc.).*

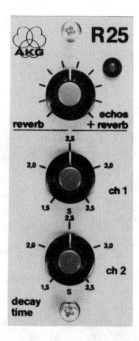

Plate Reverb Units

Plate reverb units, such as the EMT 140 TS, the first popular plate on
the market, may be still found in many studios. The EMT 140 TS
echo plate is a rectangular steel sheet $\frac{1}{64}$-inch thick and about 3 feet
high by 6 feet wide, under tension and suspended in an enclosed
frame approximately 8 feet by 4 feet by 1 foot. Reverberation is created by inducing wave motion in the plate. Another such reverb unit,
the Echoplate Chamber (Fig. 8-38), similarly creates reverberation by
inducing wave motion in a plate. These waves reach the edges of the
plate where they are reflected back across the plate to the other
edges. This process continues until the heat created through the
bending of the plate dissipates the wave energy. The waves are induced into the plate by a moving-coil driver which converts the electrical input signals into mechanical motion. The reflected waves are
picked up by contact microphones which sense the varying pressure
of the steel plate against them as the waves pass the points at which
the microphones are mounted.

Most plate units are available in a stereo version using a single
driver (mono input) and two pickup mikes which are spaced at different distances from the driver and on opposite sides of it. As a result
of this spacing, not only does the wave travelling direct from driver
to pickup take longer to reach one pickup than the other, but the
reflected waves picked up by each driver at any one point in time

**Fig. 8-38. The
Echoplate
Chamber**
*(Courtesy
Echoplate, Inc.).*

**Fig. 8-39. The
waves induced
in the plate by
the driver are
reflected at the
edges and
converted into
electrical
signals by the
pickups.**

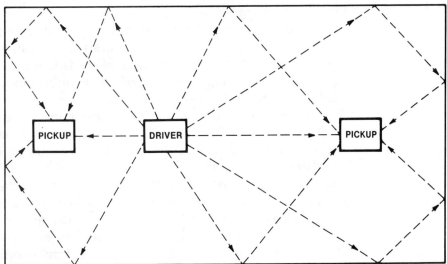

will have arrived there via different path lengths and will therefore be more or less out of phase with each other (Fig. 8-39).

The decay time of plate units is usually variable from 1 second to approximately 4 seconds. It is changed by moving a plate which is covered with acoustic absorbing material closer to the steel plate to shorten decay time and farther from the plate to lengthen decay time. The motion of the damper plate is controlled either by a handwheel on the unit or by a remote control unit.

Digital Reverb Devices

In the above methods of attaining reverberation, the user has basic control over only level, equalization, and decay time. However, a digital reverb device is a computer-based electronic effects device, which has, in many units, a great deal of control over the parameters affecting reverberation quality, such as level, decay time, pre-echo delay, low-EQ, high-EQ, variable-EQ crossover point, etc. This offers a great deal of flexibility, but is only the beginning in that the digital programs can be instantly changed to give different room and reverb type of effects. For example, a single unit is capable of creating the effect of a large concert hall, a small hall, or a bathroom, as well as plate and spring reverb simulations. It is easy to see why such units have recently become one of, if not the most, sought-after devices in today's recording studio.

Digital reverberation is accomplished by taking a signal fed to the reverb unit and performing complex digital algorithms of this signal in the form of regenerated digital delays. These delays are then repeated at a very fast rate and are programmed to follow a specifically tailored set of random-delay patterns. The random regenerations occur so quickly that the resulting effect is that of a dense reverberation. The characteristic sound of the reverb, once selected, may be tailored to the user's needs and tastes.

Among the larger studios, the most popular unit is the Lexicon 224X digital reverberator (Fig. 8-40). The 224X is comprised of two units: a reverberation/effects processor mainframe and the Lexicon Alphanumeric Remote Console (LARC). It is able to simulate the acoustics of different architectural spaces and other special effects using 18 factory preset programs with 59 reverb and effects variations, including digitally delayed chorus, resonant chords, and multiband delay. The machine also contains four split programs, enabling the unit to function as two independent full-bandwidth stereo reverbs, each with its own input. A numeric keypad addresses all programs, variations, and registers. A main display reads out information on programs, variations, and parameter values in alphanumeric form. Other displays show information on the setting sliders that can control up to 32 parameters.

**Fig. 8-40.
Lexicon 224X
digital
reverberator
with remote unit**
*(Courtesy
Lexicon Inc.).*

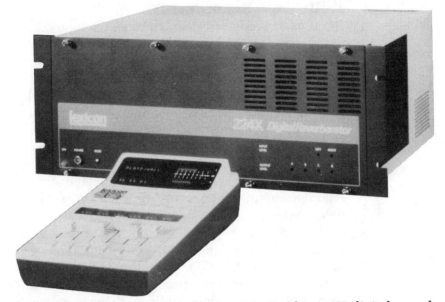

Registers, for user-created programs, in the 224X digital reverberator can store up to 36 setups, which can be off-loaded onto an audio cassette or tape machine by the LARC and then reloaded in less than one minute. This allows relocatable setups and programs to be transported to any location using a LARC-equipped 224X reverb unit.

Fig. 8-41 is a photo of the Klark-Technik DN 780 digital reverberator/processor.

**Fig. 8-41. Klark-
Technik DN 780
digital
reverberator/
processor**
*(Courtesy Klark-
Technik).*

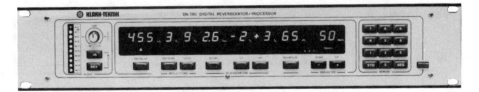

Time Delay

The next category of effects is that of *time delay*. With the advent of the digital delay unit and the digital time processor, delay has become a standard item of use in the studio; it is a part of the signature that is modern music production.

Time delay refers to the production of discrete repeats of the input signal at certain regularly spaced intervals. There are two ways of achieving this effect: using a tape recorder and using a digital delay line.

With Tape Recorder

In order to produce a discrete repeat, a tape recorder must have separate record and playback heads. The signal is recorded on the tape and played back after a short delay. The delay is equal to the distance between the record and playback heads divided by the tape speed. The playback head signal is a single repeat of what is fed to the tape recorder input (Fig. 8-42). Multiple repeats can be obtained by returning a part of the delayed output from the playback head back to the input of the recorder so that the first repeat is recorded again, delayed, and played back again (Fig. 8-43). The number of repeats that occur depends on the amount of the playback signal fed back to the recorder input. When many repeats are used, the signal gradually deteriorates into noise because each successive record and playback cycle adds a little distortion and tape noise. By returning too much playback signal to the input, a feedback type of oscillation is generated. The repeats differ from those of reverberation, regardless of how closely they are spaced, because the time intervals between them are all the same, while with an acoustic chamber, plate, or spring device, there are many different intervals between reflections. The time between repeats can be varied by either changing the spacing between the record and playback heads or by varying the tape speed.

Fig. 8-42.
Delaying a
signal using a
tape recorder.

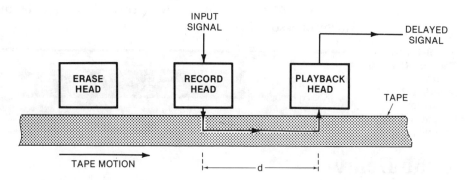

Digital Delay

The digital type of delay unit, which has become a major effects device in modern music production, produces exactly the same effect as a tape recorder used for delaying the signal, but makes the process considerably more easier (Fig. 8-44).

**Fig. 8-43.
Producing
multiple repeats
of a signal using
a tape recorder
with a single
playback head.**

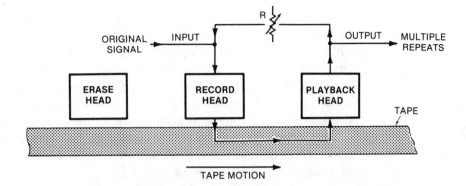

**Fig. 8-44.
Lexicon PCM 41
and PCM 42
digital delay
processor units**
*(Courtesy
Lexicon Inc.).*

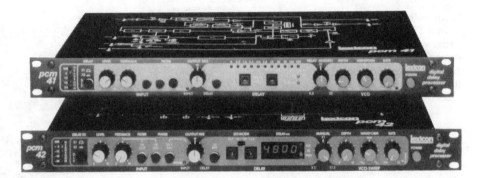

A digital delay unit encodes the analog audio signal into digital form and feeds it into a *shift register* (a storage tank for digital information). A high-frequency oscillator feeds *clock pulses* into this shift register, causing the digitally encoded number bits to move one step through the register for each pulse. The pulse frequency and the length of the shift register (the number of discrete positions between its start and its end) determine the delay time. At the output of the shift register, the signal is converted back into analog form. Multiple outputs with independently variable delay times are possible when using a single shift register by tapping the register at many different points and making the points available through a selector switch. A separate digital-to-analog converter and output amplifier is needed for each output.

Special Effects Devices

A digital special-effects device, generally based on digital delay, is a single multifunction effects device which is capable of performing a wide variety of studio effects. In recent years, these devices have become increasingly popular and useful in the control room environment. One such unit is the Lexicon Super Prime Time Model 97 programmable processor (Fig. 8-45). The Super Prime Time processor factory-preset programs include flanging, resonant flanging, doubling, tripling, chorus, slap echo, short echo, and long echo. A digital processor allows a performer to create, program, and store up to 32 special effects for two channels and 8 factory preset programs into a memory, which can be easily recalled in any sequence. Off-loading of these programs onto tape allows a large library of personal programs to be stored. The Model 97 processor comes standard with 0.96 second of full bandwidth delay, with a frequency response of 20 Hz to 20 kHz, and a dynamic range of 90 dB.

Fig. 8-45. Lexicon's Super Prime Time. A programmable digital delay processor *(Courtesy Lexicon Inc.).*

Another special-effects processor/reverberator is the Eventide Model SP 2016 signal processor (Fig. 8-46). This unit is a computer-based, versatile, general-based audio processor. The computer-controlled alphanumeric display permits controlling of all functions, input, output, I/O mix controls, and status indicators, from the front panel. Up to eleven ROM integrated-circuit software programs are available which may be interchangeably inserted into the unit. Software program features include: five different reverb programs, loop editing, multitap delay, time scramble, chorus, flanging, 4-frequency band delay, musical combs, digiplex (simulation of tape-head echo), and delay. The unit permits up to 65 sets of program and parameter presets, which can be stored and selected by the user.

Fig. 8-46. The Eventide Model SP 2016 signal processor *(Courtesy Eventide Clockworks, Inc.).*

Phasing

Phasing is a term used to describe the sound which results from a device that causes multiple sharp dips or notches in the frequency response of a system, in such a manner that the dip frequencies vary, resulting in a jet-like phasing effect (Fig. 8-47). This is caused by phase cancellations of the closely spaced signals, hence the name *phasing*.

Fig. 8-47. The phasing effect creates a comb filter with variable dip frequencies.

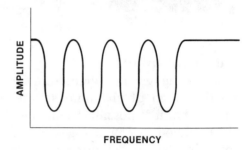

In the phasing device, the signal to be phased is fed into the input of the device where it is split into two parts. One of the parts is fed directly to a mixer, while the other one is fed first to a variable phase-shift network and then to the mixer where it is recombined with and partially cancels the original signal. A control is provided to vary the amount of phase shift produced and, thus, vary the frequencies of cancellation. The phase-shifted signal itself does not sound phased. It is the combination of the shifted signal with the unchanged signal that produces the effect. The depth of the phasing effect is controlled by the amount of cancellation (i.e., the depth of the notches), which is maximum when the levels of the direct and the phase-shifted signals are equal.

The phase-shifted signal can be used to give a stereo effect to a mono sound by putting the direct signal on one speaker and the phased signal, shifted by a constant amount, on the other speaker. Since phasing is a form of short time delay, a stereo effect is produced. If, however, the signal is later combined into mono, the phase-shifted signal will cancel certain frequencies of the direct signal and the result may be unacceptable. Therefore, when attempting this type of stereo effect, the engineer should listen to what happens when the signal is played back in mono.

One of the more elaborate phase shifters is available from Eventide Clock Works. In addition to the control of the notch frequencies, their *Instant Phaser* (Fig. 8-48) allows the depth of the phasing to be varied by changing the mixture of the direct and phase-shifted signals. Two units can also be interlocked to provide stereo phasing.

Fig. 8-48. The Eventide Clock Works Instant Phaser (*Courtesy Eventide Clock Works, Inc.*).

Pitch Changers

A pitch changer is a digital device which is capable of changing the relative pitch of an audio source without affecting its speed. This means that it is possible to change the pitch of any instrument in real time, without tricky changes in tape speed. For example, if an instrument is playing flat during a solo, it is possible to tune the track sharp during that section of music and bring the instrument back on pitch.

The most widely used pitch changer is the Eventide Model H969 Harmonizer®*. The Harmonizer® (Fig. 8-49) acts as a simultaneous digital delay unit and pitch changer, with a delay range of up to 1053 milliseconds at full bandwidth and a maximum pitch range of one octave up and three octaves down. The Harmonizer® has a set of pitch change presets, giving a precise major/minor third, fifth, or seventh octave, which are selectable either up or down.

Fig. 8-49. The Eventide Model H969 Harmonizer® (*Courtesy Eventide Inc.*).

Tape Recorder Effects

Methods of Varying Tape Speed

There are two basic methods of varying tape speed. One uses a standard tape recorder with a synchronous capstan motor (the speed of

*Harmonizer is a registered trademark of Eventide Inc.

this motor depends on the frequency of the ac power driving it), while the second method uses a *servo motor* which rotates the capstan at a speed proportional to the dc voltage applied to it.

Synchronous Motors

The first method uses a standard synchronous capstan motor, as found in older tape machines such as the Ampex AG440 or Scully 280B, and applies 110 volts ac to it at varying frequencies. The synchronous motor runs at its correct selected speed only when the power-line frequency is 60 Hz. If the frequency is raised to 65 Hz, the tape speed increases, and if the power-line frequency is lowered to 55 Hz, the tape speed slows. The power frequency is made variable through the use of a *variable-frequency oscillator (vfo)* and a power amplifier that can deliver 110 volts ac at any audio frequency. The combination of the vfo and power amplifier is called a *variable-speed oscillator (vso)*. This is diagrammed in Fig. 8-50. The standard power line (to the motor only, not to the electronics in the recorder) is replaced by the vso through the use of a switch or by removing a dummy plug on the tape transport and plugging the vso in through this socket. The speed of the tape is then varied by varying the oscillator frequency.

Fig. 8-50. Varying the speed of a hysteresis synchronous motor.

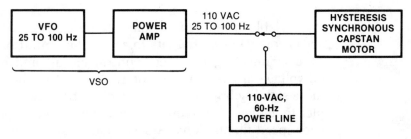

The voltage output of the power amplifier must be monitored with a meter so that neither too much nor too little voltage is applied to the motor. The voltage tolerance is from about 100 to 120 volts. Since the motor is designed to operate at 60 Hz, there is a limit as to how much the applied frequency can be varied. Normal limits are between approximately 25 and 100 Hz. The motor may tend to stall or overheat if used for long periods of time outside this range. The range of speeds attainable can be increased by using the high- and low-speed controls on the recorder in conjunction with the vso.

Servo Motor

The servo motor capstan system (Fig. 8-51) uses a dc motor which changes speed with the applied dc voltage. Normally, a tachometer pickup is mounted next to a notched wheel on the capstan or roller-idler wheel assembly which counts the number of notches that pass per second. This timed frequency is compared with the reference frequency (determined by system design) which results when the capstan rotates at the correct speed; this comparison of frequency causes an increase or decrease in the dc voltage which is applied to the motor, producing the proper pickup output frequency, thus correcting the tape speed.

Fig. 8-51. Servo motor and resolver. Slots in tach disc allow light through to sensors.

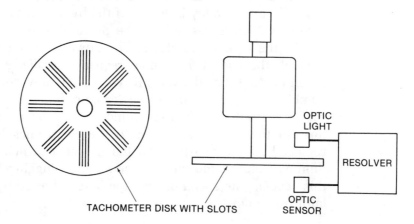

TACHOMETER DISK WITH SLOTS

OPTIC LIGHT

OPTIC SENSOR

RESOLVER

Variable speed can be achieved by varying the frequency that the speed-control circuit wants to see at the output of the tach pickup or by disconnecting the speed-control circuit and directly varying the dc voltage applied to the motor. The first of these methods provides more stability but requires more circuitry. The second method is simpler to achieve, but requires a well-regulated dc power supply to prevent speed fluctuations. A servo motor system can be made to move tape at speeds from 0 to 60 ips or higher and, therefore, it provides much more flexibility than the vso method.

Most modern mastering and multitrack recorders are supplied with a servo motor system, whether regulated by a servo capstan (open-loop recorder) or a capstanless system, where the reel motors themselves are the dc servos (zero-loop system).

Backward Effects

Another effect which can be achieved using a tape recorder is that of playing an instrument backwards. This can be achieved by recording the signal normally (Fig. 8-52) and playing it back with the tape path around the capstan and capstan idler reversed from the normal position (Fig. 8-53), or by recording the signal normally and flipping the tape upside down for playback. The second method is the one normally used in multitrack work.

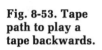

Fig. 8-52.
Normal tape
path.

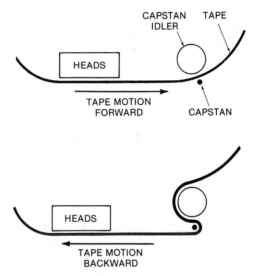

Fig. 8-53. Tape
path to play a
tape backwards.

To record a backward guitar track, for example, the master tape is placed on the machine upside down and the guitar player listens to the master tape played backward in sel sync while playing his part normally (i.e., forward). The tape is then returned to the right-side up position and the master tape plays normally while the guitar section reproduces backward.

The main concern in this procedure is that if the guitar track is to be played backward on track 2 of a 24-track tape, it must be recorded on track 23 when the tape is upside down because track 2 is the second track from the top of the head, and track 23 is the second track from the bottom. When the tape is upside down, the track order is inverted relative to the recorder heads (Fig. 8-54). The tape cannot be recorded backward, using the method of reversing the tape-threading path around the capstan to move the tape from the take-up reel to the supply reel, because the erase head would erase the signal as soon as it left the record head. Disconnecting the erase head would only result in a distorted bias signal which would be too low in level to bias the tape properly.

Backward effects can also be created by splicing in a section of tape upside down so that it is played backward. Again, the tracks would be switched around so that on a 2-track stereo tape, a sound that played forward through the left channel will play backward through the right channel.

Fig. 8-54. To achieve a backward instrument effect on track 2, flip the tape upside down and feed the instrument to channel 23 of the tape machine. This corresponds to track 2 on the upside-down tape.

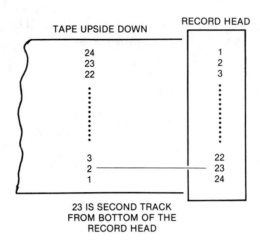

Crossfading can be used between songs on an album to eliminate the space between songs. In this, final mixes are used, and the beginning of one song and the end of another song are recorded on a multitrack tape (four tracks for stereo mixes) so that the second song begins at the desired point relative to the first song. The tape is then mixed down to two tracks so that the songs overlap smoothly. This crossfaded piece of tape is then spliced into place at the end of the first song and the beginning of the second song.

Care must be taken that the level of the songs on the mix match those of the original tapes so that the splice will not be noticeable. There is an increase in tape noise during the spliced-in portion on analog recordings, but if the tape is quiet to begin with, the increase will not be noticeable. If hiss is noticeable at the splice, both songs can be recorded entirely on multitrack tape, beginning the second song where the crossfade is desired so that the increased noise level will be constant.

Digital crossfading is accomplished at the digital editing console by transferring the two selections (using two playback machines and programming for a smooth crossfade point) to a composite digital master.

Other sound effects, such as live birds, whistles, trains, etc., can be mixed into songs by overdubbing them at the appropriate spots on the multitrack tape. Bottles, ash trays, hand claps, foot stomps, coins, or any other imaginable object can be used for effect or as rhythm instruments and even for melodic lines if they can be tuned (for example, bottles with different amounts of water in them). With proper EQ and placement in the mix, many odd noise makers can become musical instruments which complement the music but which listeners cannot identify as a specific instrument.

References

1. *AKG BX25E/BX20E Technical Bulletin*, North American Philips Corp.
2. Alexandrovich, George, "The Audio Engineer's Handbook," *db, The Sound Engineering Magazine*, Vol. 4, No. 4, April 1970, pp. 4-8.
3. Blesser, Barry, and Francis F. Lee, "An Audio Delay System Using Digital Technology," *Journal of the Audio Engineering Society*, Vol. 19, No. 5, May 1971, pp. 393-397.
4. de Gar Kulka, Leo, "Equalization . . . The Highest, Most Sustained Expression of the Recordist's Art," *Recording engineer/producer*, Vol. 3, No. 6, November/December 1972, pp. 17-23.
5. Eargle, John, and John M. Woram, "Latest Techniques and Devices in Audio Signal Processing," Panel discussion at the March 16, 1971 meeting of the New York section of the Audio Engineering Society.
6. *EMT General Catalog Specification Sheet*, March 1972, Electromesstechnik.
7. *EMT 140TS Operating Manual*, Electromesstechnik.
8. *Eventide SP 2016/H969 Product Bulletin*, Eventide Inc.
9. Factor, Richard, and Stephen Katz, "The Digital Audio Delay Line," *db, The Sound Engineering Magazine*, Vol. 6, No. 5, May 1972, pp. 18-23.
10. *Instant Phaser Specification Sheet*, Eventide Inc.
11. *Kepex Specification Sheet*, Allison Research.
12. "Lexicon 224X, Super Prime Time, PCM 41/PCM 42," *Product Bulletin*, Lexicon, Inc.
13. Nisbett, Alec, *The Technique of the Sound Studio*, New York: Hastings House Publishers, Inc., 1972.
14. "SCAMP Signal Processing," *Product Bulletin*, Audio + Designs Calrec, Inc.
15. Tremaine, Howard M., "The Audio Cyclopedia," Indianapolis: Howard W. Sams & Co., Inc., 1978.
16. *UREI 1176LN/LA-4 Specifications Sheets*, Universal Recording Electronics Industries.

9 CONSOLES

The function of a recording console is to provide control of volume, tone, blending, and spatial positioning of the signals that are applied to its inputs by microphones, electronic instruments, and tape recorders, as well as also providing a means of routing these signals quickly and reliably to the appropriate devices (tape machines or monitor systems) so that they can be heard and recorded. A console may be likened to an engineer as a pallette is likened to an artist; it is a place at which to experiment, blend, and control all the possible variables of sound. The console allows the subtle combination and mixture of electrical information, which would otherwise be rigidly fixed and uncontrolled.

Before the introduction of multitrack tape machines, all the sounds that were to be part of a recording were mixed together, at one time, during a live performance. If the recorded blend or *mix*, as it is called, was not satisfactory or if one musician made a mistake, the selection had to be performed again and again until the desired balance and performance was obtained. With the availability of multitrack recording, the production of a modern recording has become a process involving three stages: recording, overdubbing, and mixdown.

Recording

During the first stage, called *recording*, one of two methods will be followed. Either all the musical instruments being used in the song are recorded onto tape in one live pass or just a few of the musicians will play and record their parts, thereby forming the instrumental foundation upon which the song will be built. The latter procedure, which has come to be the most accepted fashion for popular music, has come to be known as the *laying down* of basic or rhythm tracks. *Basic tracks* consist of those instruments of a song which provide the rhythm foundations of a song and are, most commonly, the drums, bass and rhythm guitars, and a piano, with an optional guide vocal. The instruments are divided into groups of one or more, as determined by the desire of the producer and the engineer in exercising individual control over their tone, level, and placement in the final product.

In general, each instrument of a popular recording project is recorded onto a separate track of the master tape. This is done by connecting the microphones to the *input section* of the console and then assigning each microphone to a console output through a series of switches or push buttons called the *channel assignment matrix*; each console output is connected to a different track on the multitrack recorder. If several instruments are to be recorded onto one track, their microphones are connected to the same console output channel and the balance between each instrument is adjusted by using the level control provided for each microphone input. Although tonal and placement changes can be made at this point, the bulk of these corrections are made at a later stage. The signal recorded onto the multitrack tape is recorded at a level as high as possible, but without overloading the tape and also without regard to the musical relationship of the instruments on the other tracks. This is done to achieve the best signal-to-noise ratio possible on each track of the tape, so that the final product will not be impaired by the audibility of tape hiss.

The recording stage is of vital importance to the outcome of the final recording. The rhythm tracks are most often the driving backbone of a song, and to record them improperly onto a tape is just asking for a recording that is, at best, of unacceptable standards. Getting the sounds onto tape right, without having to rely upon the "We'll fix it in the mix" method, will get you off to a start that will not be regretted later.

Monitoring

Since the producer and engineer must be able to hear the instruments as they are recorded and played back, and because the levels are recorded irrespective of the overall musical balance, the console and tape machine outputs are connected to a *monitor mixer*, which is used to judge the quality of the performance. In the *program* (record) mode, the console outputs (the microphone signals) are fed to the monitor mixer where the different groups of instruments are blended and the output is then fed to the monitor speakers while being recorded separately on the master tape. In order to hear an approximation of the final product, a separate monitor level is provided for each console output, as well as left/right panning for signal placement over the monitor speakers. The monitor mixer also has echo facilities so that the monitor output can be even closer to the final product. *Cue* or *foldback* monitor levels adjust the monitor balance levels for the musicians' headphone "mixes" in much the same way as the studio monitor "mix" but they are totally independent. In fact, there may be two or more separate cue "mixes" available to the musicians, depending upon what their listening needs are.

In the *tape playback* mode, the tape machine outputs are connected to the inputs of the same monitor controls, at the same levels; this permits the monitor and cue balances and the placements which were set during recording to be heard again during playback without readjusting the level controls.

Overdubbing

Instruments not present during the original performance may be added onto the existing multitrack tape in a second stage process called *overdubbing*. In this stage, musicians listen to the previously recorded tracks over headphones and play along with these tracks. If one or more musicians have made minor mistakes during an otherwise good performance or if additional instruments are needed to be added to the initial basic tracks in order to finish the recording project, the new performances are recorded in synchronization with the original performances onto unrecorded tracks or onto tracks containing information which is no longer desired. Generally, monitor switching is provided at the multitrack tape recorder position, with the monitor input being fed by tape playback off of the recording head, while the monitor track, which is in the record mode, is fed off of the console input. This allows the engineer to listen to the entire program in sync.

Mixdown

After all the desired musical parts have been performed and recorded to the satisfaction of the producer and the engineer, the *mixdown* or *remix* stage takes place. At this point, the inputs to the console are fed by the playback outputs of the multitrack recorder. This is often accomplished by switching the console to the *mixdown* mode or by changing the microphone/line switches, on the appropriate console inputs, to the *line* position.

The master tape is then played repeatedly while adjustments to the level, equalization, effects, reverberation, and panning are made on each track. At this time, the individual signals are blended into a composite stereo or mono signal which is sent to the console outputs. The outputs are then routed to a 2-track master recorder where the final mixed product is recorded. When approved, this recording is called the *final mix*. It will be made into the final product, be that an album, cassette, tv or radio commercial, or the sound track for the latest hit video. The mix tape can be played back through the monitor system by switching the console into the tape playback mode; this connects the monitor speakers to the outputs of the mixdown recorder.

Professional Consoles

The majority of consoles used in modern professional multitrack studios have similar controls and capabilities (Figs. 9-1 and 9-2). They differ mainly in appearance, in location of controls, in methods of information storage for automation, and, on the newer top-line consoles, in their methods of signal routing by computer. Modern consoles are usually designed using plug-in input strip modules, called I/O modules, which are mounted in custom-designed enclosures (Figs. 9-3 and 9-4).

I/O stands for input/output. The module is so named because all of the necessary electronics, including the microphone preamp and the signal routing and track assignments, are all located on a single I/O strip. I/O modules permit modular construction of consoles which then can be configured to fit the needs of the studio; this allows for expansion where necessary. The plug-in nature of the strips makes them easily removable for servicing and interchangeability.

Patch Bays

The components of the console interconnect at the *patch bay* (also called a *patch panel* or *patch rack*). A typical patch bay is shown in Fig. 9-5.

A patch bay is a panel which contains a jack corresponding to the input and output of every discrete component or group of wires in the control room. One jack is said to be *normalled* to another if the components connected to the two jacks are connected directly to each other when there is a plug in neither jack (Fig. 9-6), but are not connected if a plug is inserted. When a plug is inserted into one or both of the two jacks, it is said that the normal between those jacks has been *broken* (Fig. 9-7).

Fig. 9-1. MCI JH-636 AF/LM console *(Courtesy MCI, a Division of Sony Corporation of America).*

The purpose of breaking normalled connections, by the insertion of a plug, is to enable the engineer to use *patch cords* to connect different or additional pieces of equipment between two normally connected components. For example, a limiter can be temporarily patched between a mike preamp output and an equalizer input. The same limiter could be later patched between a tape machine output and a console line input. Other uses of the patch bay are to bypass defective components or to change a signal path in order to achieve a certain effect. Patch bay jacks are usually balanced tip-ring-sleeve types having two conductors plus ground. They are used with balanced circuits (Fig. 9-8) and come in two standard sizes: military telephone and tiny telephone. Unbalanced jacks and plugs have only tip-and-sleeve connectors (Fig. 9-9).

Preamps

The output level of a microphone is very low; thus, it requires a specially designed amplifier, called a *mike preamp*, which raises the low-level output to a level required by the console so that the signal can be further processed without degrading the *signal-to-noise ratio*.

**Fig. 9-2. NEVE
8128 console
with NECAM 96
automation
system** *(Courtesy
Rupert Neve
Inc.).*

The electrical signals produced by a mike are actually electrons in
motion. Since, in any conductor at temperatures above absolute zero
(− 273 °C), electrons are moving from atom to atom, there is a limit
to how quiet an electrical system, with a certain frequency response,
can be.

Fig. 9-3. MCI JH600 I/O module
(Courtesy MCI, a Division of Sony Corporation of America).

LINE SWITCH (three position) selects a boost/cut of the line input signal of +6dB, 0dB, or −6dB.

MIC GAIN SWITCH and TRIM POT; with switch up trims mic gain by 12 to 35dB; with switch down trims mic gain by 30 to 65dB.

PHANTOM POWER SWITCH turns microphone phantom power on when pressed down.

LINE SWITCH when up selects microphone input as channel source; when down selects line input as channel source.

BUS PAN POT pans output between odd and even channel buses assigned by channel buttons.

ODD/PAN/EVEN SWITCH (three position) allows panning with Pan Pot or selects only odd or even channel buses.

DUMP SWITCH when pressed assigns the Monitor 2-Mix output to the assigned channel buses.

HIGH EQ LEVEL POT boosts or cuts the high frequencies between ±14dB at the frequency selected.

MID EQ LEVEL POT boosts or cuts the mid range frequencies from +14dB or −14dB.

LOW EQ LEVEL POT boosts or cuts the low frequencies from +14dB to −14dB.

EQ IN SWITCH switches the EQ circuitry when pressed in; when out EQ circuitry is bypassed.

PHASE REVERSE SWITCH reverses the phase of the signal through the EQ section.

EQ CHANNEL SWITCH when out EQ is in monitor circuit; when in EQ is in channel circuit.

PRE SWITCH selects pre-monitor fader input to Send 1-2 when in; when out input is post fader.

WET SWITCH connects channel bus output to record echo on multitrack.

PRE SWITCH selects pre-monitor fader input to Send 3-4.

PRE SWITCH selects pre-monitor fader input to Send 5-6.

2-MIX selects 2-Mix monitor output as input to Send 5-6.

MUTE SWITCH mutes the 2-Mix monitor output.

BUS SWITCH routes output of monitor fader to input of channel fader.

VCA REVERSE SWITCH places VCA fader into channel and rotary fader into the monitor.

LEVEL POT rotary fader controls channel level (unless VCA is pressed).

REVERSE SWITCH switches monitor input from line input to channel line output.

PRE-FADER LISTEN SWITCH routes monitor fader input to PFL ACN bus.

AUTO NULL LEDs indicate which direction to move fader in update and rewrite modes and when reassigning master.

FADER WRITE BUTTON places channel (or group) into write or rewrite mode (LED on). If LED flashes, fader must be nulled to enter or leave write or rewrite mode.

FADER UPDATE BUTTON places channel (or group) into update mode (LED on). If LED flashes, fader must be nulled to leave update mode.

GROUP SELECT SWITCH (ten position) assigns fader to any of eight groups or to local (L) control.

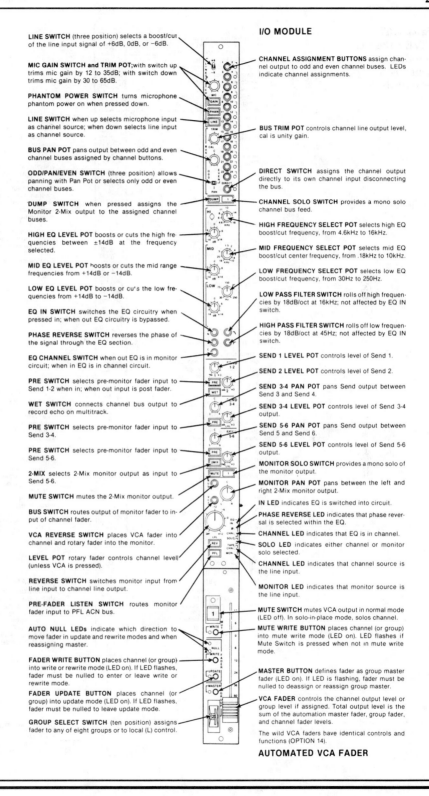

I/O MODULE

CHANNEL ASSIGNMENT BUTTONS assign channel output to odd and even channel buses. LEDs indicate channel assignments.

BUS TRIM POT controls channel line output level, cal is unity gain.

DIRECT SWITCH assigns the channel output directly to its own channel input disconnecting the bus.

CHANNEL SOLO SWITCH provides a mono solo channel bus feed.

HIGH FREQUENCY SELECT POT selects high EQ boost/cut frequency, from 4.6kHz to 16kHz.

MID FREQUENCY SELECT POT selects mid EQ boost/cut center frequency, from .18kHz to 10kHz.

LOW FREQUENCY SELECT POT selects low EQ boost/cut frequency, from 30Hz to 250Hz.

LOW PASS FILTER SWITCH rolls off high frequencies by 18dB/oct at 16kHz; not affected by EQ IN switch.

HIGH PASS FILTER SWITCH rolls off low frequencies by 18dB/oct at 45Hz; not affected by EQ IN switch.

SEND 1 LEVEL POT controls level of Send 1.

SEND 2 LEVEL POT controls level of Send 2.

SEND 3-4 PAN POT pans Send output between Send 3 and Send 4.

SEND 3-4 LEVEL POT controls level of Send 3-4 output.

SEND 5-6 PAN POT pans Send output between Send 5 and Send 6.

SEND 5-6 LEVEL POT controls level of Send 5-6 output.

MONITOR SOLO SWITCH provides a mono solo of the monitor output.

MONITOR PAN POT pans between the left and right 2-Mix monitor output.

IN LED indicates EQ is switched into circuit.

PHASE REVERSE LED indicates that phase reversal is selected within the EQ.

CHANNEL LED indicates that EQ is in channel.

SOLO LED indicates either channel or monitor solo selected.

CHANNEL LED indicates that channel source is the line input.

MONITOR LED indicates that monitor source is the line input.

MUTE SWITCH mutes VCA output in normal mode (LED off). In solo-in-place mode, solos channel.

MUTE WRITE BUTTON places channel (or group) into mute write mode (LED on). LED flashes if Mute Switch is pressed when not in mute write mode.

MASTER BUTTON defines fader as group master fader (LED on). If LED is flashing, fader must be nulled to deassign or reassign group master.

VCA FADER controls the channel output level or group level if assigned. Total output level is the sum of the automation master fader, group fader, and channel fader levels.

The wild VCA faders have identical controls and functions (OPTION 14).

AUTOMATED VCA FADER

Fig. 9-4. The NEVE 8128 I/O module
(Courtesy Rupert Neve Inc.).

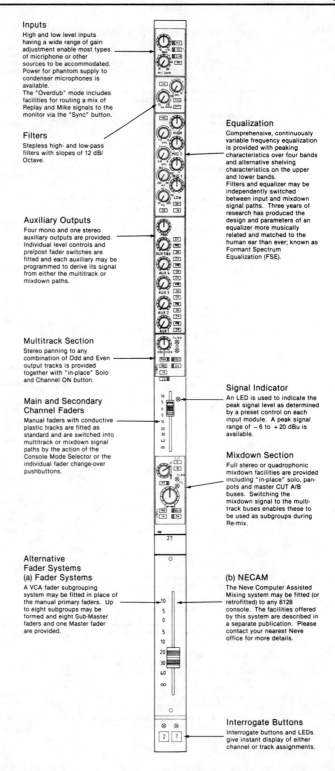

Inputs

High and low level inputs having a wide range of gain adjustment enable most types of micriphone or other sources to be accommodated. Power for phantom supply to condenser microphones is available.
The "Overdub" mode includes facilities for routing a mix of Replay and Mike signals to the monitor via the "Sync" button.

Filters

Stepless high- and low-pass filters with slopes of 12 dB/ Octave.

Auxiliary Outputs

Four mono and one stereo auxiliary outputs are provided. Individual level controls and pre/post fader switches are fitted and each auxiliary may be programmed to derive its signal from either the multitrack or mixdown paths.

Multitrack Section

Stereo panning to any combination of Odd and Even output tracks is provided together with "in-place" Solo and Channel ON button.

Main and Secondary Channel Faders

Manual faders with conductive plastic tracks are fitted as standard and are switched into multitrack or mixdown signal paths by the action of the Console Mode Selector or the individual fader change-over pushbuttons.

Alternative Fader Systems
(a) Fader Systems

A VCA fader subgrouping system may be fitted in place of the manual primary faders. Up to eight subgroups may be formed and eight Sub-Master faders and one Master fader are provided.

Equalization

Comprehensive, continuously variable frequency equalization is provided with peaking characteristics over four bands and alternative shelving characteristics on the upper and lower bands.
Filters and equalizer may be independently switched between input and mixdown signal paths. Three years of research has produced the design and parameters of an equalizer more musically related and matched to the human ear than ever; known as Formant Spectrum Equalization (FSE).

Signal Indicator

An LED is used to indicate the peak signal level as determined by a preset control on each input module. A peak signal range of − 6 to + 20 dBu is available.

Mixdown Section

Full stereo or quadrophonic mixdown facilities are provided including "in-place" solo, pan- pots and master CUT A/B buses. Switching the mixdown signal to the multi- track buses enables these to be used as subgroups during Re-mix.

(b) NECAM

The Neve Computer Assisted Mixing system may be fitted (or retrofitted) to any 8128 console. The facilities offered by this system are described in a separate publication. Please contact your nearest Neve office for more details.

Interrogate Buttons

Interrogate buttons and LEDs give instant display of either channel or track assignments.

**Fig. 9-5. A patch
bay.**

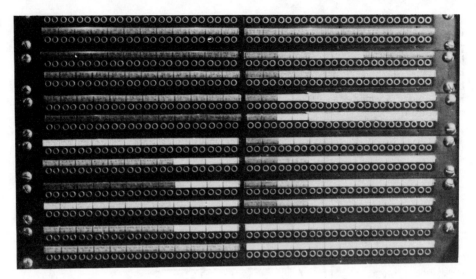

**Fig. 9-6.
Normalled
jacks; the
preamp output
is normalled to
the EQ input.**

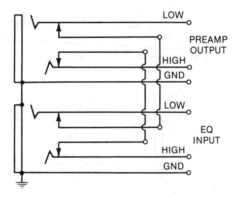

When any signal is amplified, noise caused by random electron
motion in the signal source is also amplified. Transistors and vacuum tubes also generate noise which is added to the signal. Since
amplifiers cannot distinguish between desired signals and noise,
both are amplified equally. Once the signal reaches a certain level,
however, the noise added to it by high-quality components is insignificant. The signal-to-noise ratio is primarily determined by the
mike preamp which raises the signal to this higher level. Balanced
mike lines and mike preamp inputs reduce the noise content of the
signal, in comparison to unbalanced lines and inputs, by cancelling
out "hum" and the outside noises that are induced into the mike
cords.

Fig. 9-7. A plug inserted into the preamp output jack or the EQ input jack breaks the normalled connection.

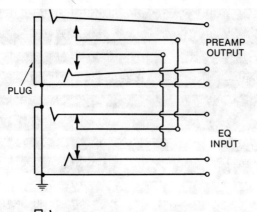

(A) Plug in preamp output jack.

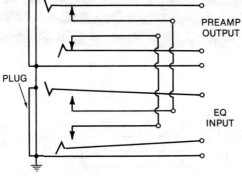

(B) Plug in EQ input jack.

Fig. 9-8. Types of balanced jacks and the plug used with them.

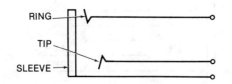

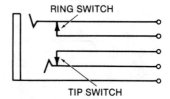

(A) Balanced jack.   *(B) Balanced jack with switches.*

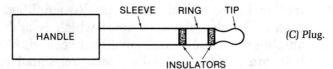

(C) Plug.

Fig. 9-9. Unbalanced jack and plug.

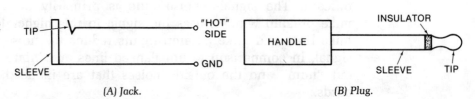

(A) Jack.   *(B) Plug.*

All amplifiers have a maximum undistorted output level. If a signal which is fed into a mike preamp is high enough so that the preamp gain raises it above this level, the signal becomes distorted. Fig. 9-10 shows the input/output curve for a mike preamp that has 40-dB gain and a +30-dBm maximum output before clipping. The preamp follows the 45° undistorted linear line when the input signal is below −10 dBm. When the input signal rises above this level, the gain of the amp tries to raise the output signal above +30 dBm, so the signal is clipped, producing severe distortion. To avoid this, a microphone preamp gain trim provides a means of reducing both the gain of the preamp and the level of the incoming signal. This mike trim control varies the preamp gain over an approximate 40-dB range and allows the gain to be set so that the input signal is amplified only as much as is needed by the console to obtain the lowest noise levels.

Fig. 9-10. The input/output curve for a microphone preamplifier.

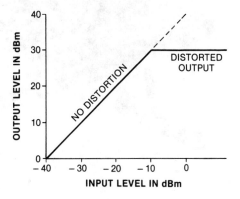

The output of the mike preamp is often fed to a phase reversal switch. This switch is used to effect a 180° change of phase in the input signal to compensate for any microphone or mike cable that is out of phase. This prevents cancellation which occurs when out-of-phase mikes are mixed.

Equalizers

The output of the EQ switch (when present) is fed to the equalizer of the input strip. Equalizers provide simultaneous control over three overlapping frequency ranges and, on newer models, often involve parametrics, thus allowing continuous control over frequency, boost, and, possibly, over the bandwidth.

The high frequencies on the MCI JH-600 Series consoles (Fig. 9-11), for example, sweep between 4.6 and 16 kHz, while the midrange frequencies range between 180 Hz and 10 kHz and the low frequencies range from 30 Hz to 250 Hz. All three ranges produce peaking and dipping curves but the high and low ranges can also be individually switched to produce peaking or shelving curves. Often low and high bandpass filters are present which filter out any signals that are above or below the audio range (such as high-frequency oscillations or rumbles which are transmitted through the floor or studio walls). These are attenuated with minimal or no effect on the sound of the program material. The EQ channel switch (Fig. 9-12) allows the engineer to either place the selected EQ in circuit or remove it, without any switching noise.

Fig. 9-11. The JH-600 Series audio console *(Courtesy MCI, a Division of Sony Corporation of America).*

The equalizer output follows basically four routes (Fig. 9-13):

1. To the auxiliary sends.
2. To the record monitor input (which precedes the channel output fader).
3. To the channel assignment matrix.
4. To the console output buses (which follow the channel output fader levels).

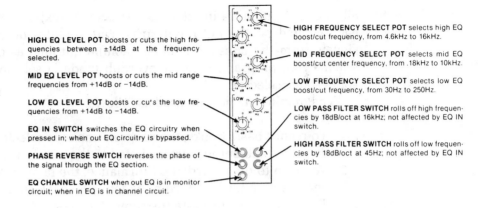

**Fig. 9-12.
Equalizer
section of I/O
module for the
JH-600 Series
audio console**
*(Courtesy MCI, a
Division of Sony
Corporation of
America).*

HIGH EQ LEVEL POT boosts or cuts the high frequencies between ±14dB at the frequency selected.

MID EQ LEVEL POT boosts or cuts the mid range frequencies from +14dB or −14dB.

LOW EQ LEVEL POT boosts or cuts the low frequencies from +14dB to −14dB.

EQ IN SWITCH switches the EQ circuitry when pressed in; when out EQ circuitry is bypassed.

PHASE REVERSE SWITCH reverses the phase of the signal through the EQ section.

EQ CHANNEL SWITCH when out EQ is in monitor circuit; when in EQ is in channel circuit.

HIGH FREQUENCY SELECT POT selects high EQ boost/cut frequency, from 4.6kHz to 16kHz.

MID FREQUENCY SELECT POT selects mid EQ boost/cut center frequency, from .18kHz to 10kHz.

LOW FREQUENCY SELECT POT selects low EQ boost/cut frequency, from 30Hz to 250Hz.

LOW PASS FILTER SWITCH rolls off high frequencies by 18dB/oct at 16kHz; not affected by EQ IN switch.

HIGH PASS FILTER SWITCH rolls off low frequencies by 18dB/oct at 45Hz; not affected by EQ IN switch.

**Fig. 9-13.
Flowchart of an
I/O module.**

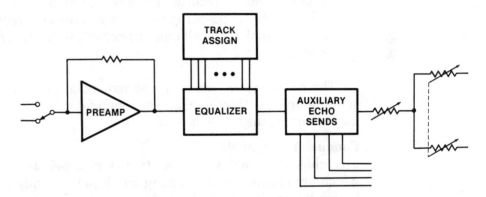

Auxiliary Sends

The auxiliary sends (aux sends) are the *effects sends* of the console. There may be up to eight sends on some console models. These are simply volume trim controls which are mixed together to form a single output or are combined to form a stereo output; these outputs can then be sent to whichever device will give the desired effect.

This section can be looked at as simply an output mixer that has as many inputs as are available on the console. Any or all channels can be sent to this output. If there are 4, 6, or 8 aux sends available, the engineer could have that many "mixes" driving the "effects" devices. On more recent console designs, the "reverb" or "echo" sends will most likely be auxiliary sends which are delegated for that purpose.

Record Monitor Input

The channel signal is sent to the monitor input such that, in the record mode, the instrument being recorded will be fed into the speaker and cue monitor mix. The multitrack monitor mixer generally consists of a 24-input (one for each track) monitor submixer, each located on the I/O input strip. The modules derive their input signals from the console or tape machine outputs, as determined by whether the program, tape, or overdub mode is selected on the console's status selector.

The monitor modes are used as follows:

1. To record basic tracks, the program mode is selected and the console track outputs are assigned to the monitor.
2. To listen to the recorded performance, the tape mode is selected and the monitor is fed from the tape playback signals.
3. To overdub one or more instruments, the sync button is selected, and the recorder outputs are fed to the monitor section. The tracks being recorded will feed the recorder's output signal, while all other tracks will feed the sync playback signal.

The monitor mix can feed a stereo signal to the control room or studio monitors. One or more separate cue mixes in mono or stereo may be sent to the studio headphones.

Channel Assignment

The channel signal is routed to the channel assignment buttons which are capable of distributing any input into any or all tracks on the multitrack recorder. Thus, for example, if a synthesizer plugged into input No. 14 is being recorded onto track No. 14, the engineer will simply press the button marked 14. However, if a second overdub is needed, all he has to do is punch out button 14 and push button 15, thus sending the output to track No. 15 while still using input strip No. 14. This can seem easy until you realize that while you are recording on input No. 14, your monitor level is coming up on input No. 15. It is just a matter of keeping track of where you are to avoid confusion.

Console Output Buses

The fourth and final "send" of consequence is the one which is sent to the mixdown buses or outputs of the console. These buses/outputs will generally be a stereo mixdown pair whose output is sent to the 2-track mixdown recorder. Each input strip will contain an associated *output fader* (which determines the level) and the *pan pot* (which determines the placement of the instrument in the stereo field). All of the strip output signals are then combined with the auxiliary and echo returns and are mixed onto one stereo output bus.

Metering

The output of each of the twenty-four or more line output amplifiers (found on major consoles) is monitored by a meter indicator which displays the waveform excursions, indicating the level of signal being recorded onto tape. If the signal level is too low, tape noise will be a problem when the recording is played back. If the level is too high, the tape and, possibly, the console or tape recorder amplifiers may cause distortion of the signal. Proper recording level is achieved when the highest reading on the meter is near the zero (0) level, although levels slightly above or below this will not cause difficulties (Fig. 9-14).

Fig. 9-14. VU meter readings.

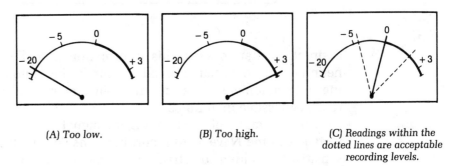

(A) Too low.   (B) Too high.   (C) Readings within the dotted lines are acceptable recording levels.

Multitrack console metering bridges are available in basically two types (Fig. 9-15): a VU meter bridge and an LED or light-bar meter display. The VU bridges incorporate as many VU meters as there are tracks and display both the console track output level in the record mode and the tape return level in the playback or overdub mode. The main stereo output buses, as well as all the aux send outputs, are metered. LED or light-bar displays, on larger consoles, often have the advantage of being switchable from *rms* to *peak* level readings.

Digital Signal Routing and Processing

In recent years, a new breed of console has begun to emerge on the market. These consoles are using digital technology to route signals from one console section to another with the aid of a computer processor and a keyboard interface. Track assignment, auxiliary send and return routing, channel on/off, EQ in/out, as well as a multitude of digitally switched functions, may be assigned from one central keyboard. This feature allows for the elimination of many of the discrete switches normally required for track selection and many other functions—switches which tend to be costly and potentially faulty. Since all the just-mentioned functions are digitally encoded and stored, it is

Fig. 9-15. A
plasma and VU
meter display
for the JH-600
console.

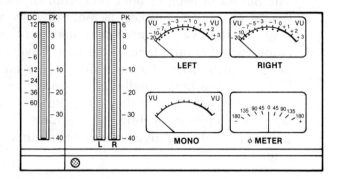

a simple matter to store this information onto floppy disk or magnetic tape for automated recall. In combining automated signal routing with automated level controls, an almost total session/mixdown
recall is possible in minutes.

One example of such a digital signal router is that which is
installed in the Neve 8128 recording console (Fig. 9-16). Using touch-
sensitive switching, the track assignment panel of the 8128 console
allows up to 56 input channels to be routed to any or all of the output
tracks. The internal solid-state memory can store up to four complete
track assignments with total recall. Auxiliary outputs 1 through 4
may be assigned to any or all of four cue or reverb "sends" with the
information being stored in memory. The provision for an *assign lock*
is used to prevent accidental erasure of these storage functions.

Rupert Neve Inc. has also introduced its new DSP (Digital Signal
Processing) console onto the market (Fig. 9-17). The DSP console is
one of the first of its kind, in that it is a totally digital console. Not
only is signal routing accomplished digitally, but the audio signal
path itself is digitally encoded. With the advent of total digital signal
processing, many new options and design features present themselves. One such feature, beyond total signal routing and signal-path
designations, is the ability to incorporate assignable multifunction
controls. These are provided on each DSP console fader. These single sets of controls can be switched to a variety of different functions
at will, so that cue sends for all channels, for example, can be displayed across the console at one moment and then the controls are
reassigned to display reverb send levels the next. This allows for
many of the console's controls to be "doubled up," thereby eliminating virtually hundreds of discrete controls and duplicated devices

Fig. 9-16. Neve 8128 panel *(Courtesy Rupert Neve Inc.).*

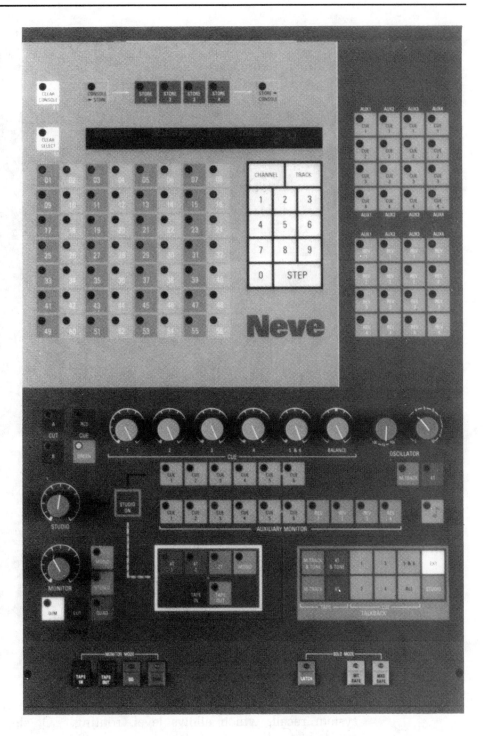

(Fig. 9-18). The DSP panel while incorporating digital processing of dynamics control, EQ, and filtering, also incorporates an integral digital delay which is assignable to any or all input or track assignments, thus allowing versatile control over time.

Fig. 9-17. DSP console
(Courtesy Rupert Neve Inc.).

Fig. 9-18. DSP input and aux control panel
(Courtesy Rupert Neve Inc.).

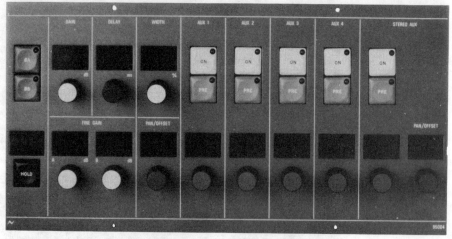

Riding over all of these features is the advantage of absolute total system recall, which allows level, routing, EQ, signal processing, etc., to be totally automated.

Optimizing Console Noise

Typical output levels for dynamic and ribbon microphones are on the order of −55 dBm (re: 10 microbars) and −40 dBm for condenser microphones. These levels are too low to be sent directly to the recording console or the tape machine, so specially designed microphone preamps are used to raise these levels. These preamps must have a very low equivalent input noise (e.i.n.) figure. The lower this e.i.n. figure, the less noise will be added to the signal before it is amplified and, since both signal and noise are amplified, the lower the noise that will be in the final signal. The signal-to-noise (s/n) ratio is the difference in dB between the level of the signal and the level of the noise. For example, a console may use a mike preamp which has an equivalent noise of −127 dBm; an input signal of −55 dBm fed into the preamp will yield the difference between the −55-dBm signal and the −127-dBm noise, or a signal-to-noise ratio of 72 dB. If a mike pad is used before the preamp input to decrease the mike signal to −65 dBm, the maximum s/n ratio would be only 62 dB. Therefore, pads should only be used when they are needed to prevent mike preamp overload distortion.

The s/n ratio at the preamp output can be degraded through poor console operating procedures. As the signal flows through the console, it passes through several stages of attenuation, each of which is followed by amplification. As long as the signal level stays high enough so that the equivalent input noise of the next amplifier is far enough below the level of the noise component already present in the signal (so that its contribution to the amplifier output is insignificant), attenuation and reamplification does not degrade the signal-to-noise ratio. The maximum amount of attenuation in dB that can be inserted between the output of amplifier No.1 and the input of amplifier No. 2, and still cause less than a 0.1-dB reduction in the s/n ratio, can be computed by the following formula:

$$dB = \text{Gain 1} - 20\,dB - (\text{e.i.n. 2} - \text{e.i.n. 1}) \qquad \textbf{(Eq. 9-1)}$$

As attenuation is increased beyond this amount, both the signal and the noise level fed to amplifier 2 decrease, but the equivalent input noise of amplifier 2 does not. Therefore, e.i.n. 2 becomes more and more a significant portion of the amplifier input signal. For example, beginning with a 72-dB s/n ratio that is composed of −10-dBm signal and −82-dBm noise at the output of a 45-dB-gain mike preamp, and attenuating this by 25 dB produces a −35-dBm signal and a − 107-dBm noise level. If the equivalent input noise of the next amplifier is equivalent to the mike preamp (− 127 dBm), this is a permissible amount of attenuation and the e.i.n. of the next stage does not significantly contribute to the noise level.

If the gain of the next stage is set to 25 dB, the signal and noise level are bought back up to − 10 dBm and − 82 dBm, respectively. The signal-to-noise ratio is unchanged throughout the various level changes, remaining at 72 dB (Fig. 9-19). Starting with a mike output level of − 55 dBm and an equivalent noise in the mike preamp of − 127 dBm, there is an s/n ratio of 72 dB. The mike preamp raises the level of both the signal and the noise by a figure of 45 dB, a fader attenuates them by 25 dB, and a booster amp raises them by 25 dB. The 72-dB ratio is maintained throughout. Noise added by the input stage of the booster amp is lower than the noise level of the preamp output and can be disregarded. If, on the other hand, 50 dB of attenuation was used after the microphone preamp, the signal would be reduced in level to − 60 dBm, while the noise level would fall to only − 130.9 dBm (the minimum noise level possible in a circuit with a frequency response of 20 Hz to 20 kHz) rather than to − 132 dBm. This noise level is added to the − 127 dBm e.i.n. of the next stage, producing a total noise level at its input of approximately − 125.5 dBm. To obtain the original signal level of − 10 dBm, the next amplifier must have 50 dB of gain. This raises the noise level up to − 75.5 dBm, resulting in an s/n ratio of only 65.5 dB (Fig. 9-20).

Fig. 9-19. Good console operating procedures do not degrade the signal-to-noise ratio.

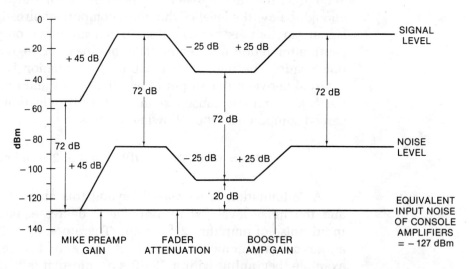

Attenuating the input signal within 20 dB of the amount of the mike preamp gain lowers the s/n ratio. The fader attenuates the input

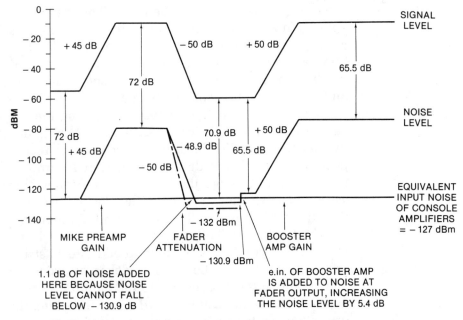

**Fig. 9-20.
Attenuating the
input signal
within 20 dB of
the amount of
the mike preamp
gain lowers the
signal-to-noise
ratio.**

signal by 50 dB to − 60 dBm and reduces the noise generated by the mike to the theoretical − 130.9 limit. The − 127 dBm e.i.n. of the booster amp is added to the − 130.9 dBm noise at the fader output, producing a − 125.5-dBm noise level at the input of the booster amp, for an additional 5.4 dB of noise. The ratio at the input of the booster amplifier is a − 60-dBm signal to a − 125.5-dBm noise, or 65.5 dB; this is a total noise increase of 6.5 dB. There has been a 6.5 dB of signal-to-noise loss because the signal was attenuated too much. By being aware of the gain and the e.i.n. figures of the amplifiers in the console, the engineer can optimize the levels within the console stages for maximum signal-to-noise.

Multitrack Recording

During a multitrack recording session, the engineer may encounter any number of possible situations. Since most tape tracks are fed by a single microphone, only a few mixing channels may be in use at once during an overdub, but the next session could require all the channels and just a bit of ingenuity to make it all fall into place. This is where you realize that the key to good console operation is reliability and flexibility.

With an input being in either the mike or line position, the first duty at hand is to set the preamp gain trim for an optimum input gain setting. Once the proper level and phase settings have been made, EQ adjustments may be made. The idea of EQ is tackled differently by many engineers. Many will, at this point, adjust the EQ settings to get the desired sound onto tape, while others prefer to make these adjustments later in the mixdown stage. Still others prefer to use little or no equalization, relying solely upon microphone choice and placement to achieve the desired sound. The most accepted method is a combination of the two, getting the best sound that you can through mike choice and placement and, then, correcting for deficiencies with EQ. With the console in the record mode, it is then possible to send this dry signal (a signal without added effects or reverberation) through the input's master fader, through the channel assign matrix, and straight to the desired track on the multitrack recorder. This signal will be also fed to the monitor section, allowing it to be heard over both the studio monitors and the headphone cue mix. At this point, it is possible to add effects strictly in the monitor section, such as reverberation or delay, by using the auxiliary sends and assigning the aux returns to the monitor section. In this way, the approximate sound can be heard while recording the signal onto tape dry, allowing for final judgements later in the mixdown. During a recording session, quite often there arises a need to listen to a single instrument or track in a complicated mix in order to make fine adjustments. For this reason, the *solo* or *pre-fade listen (pfl)* was incorporated. The solo button allows listening to the desired track while muting all the other tracks over the monitor. The solo function in no way affects the recorded signal, or the cue signal which is sent to the studio headphones. It is strictly a studio monitor function. Some consoles are fitted with a mute button which does the reverse; it allows all other signals to be heard while muting the selected signal.

In most moderate to top-line consoles, the *overdubbing process* is quite straightforward. By placing the console into the overdub (O/D) mode, the monitor inputs are electronically switched by the multitrack recorder; thus, if an overdub is to be made onto track No. 9, it is only necessary to place the multitrack recorder into the sync mode and depress the record button on track 9 at the desired place on the tape. The monitor section will reproduce, in sync, all tracks but number 9, while allowing us to hear the signal being recorded.

In the *mixdown mode*, the playback outputs of the multitrack recorder are fed to the line inputs of the console. The input strip's signal flow allows for both total control and flexibility for each separate track over the volume, equalization, effects, reverberation, and placement within the stereo field. The signal will flow from the preamp to the EQ section and, then, from this point, the signal is split between the input strip's main fader and the aux sends, where the effects may be mixed. The main faders are split between the two stereo output buses by the pan pot. The signals from all the active input strips are then combined into signals at the output buses, along with the echo and aux returns. The composite signals are then sent to the stereo mixdown machine for recording.

The studio *talk-back button* and *level control* permit the engineer and producer to speak to people in the studio over the studio loudspeakers or the musicians' headphones. The control room *talk-back mike* is mounted in the console and has a gain trim control so that its sensitivity can be adjusted to that which is needed for good communication. In addition, a *slate button* and *level control* feed the talkback mike output to the console output buses for recording onto tape in order to identify different selections or "takes" of the same selection. A low-frequency tone (20 or 30 kHz) may be recorded onto tape along with the *slate identifier*. This signal, when fast forwarded on tape, is raised in frequency to the center of the audio spectrum where it stands out clearly. Since both human hearing and most monitor speakers are less sensitive at very low frequencies, the playback of the tone at normal speed will not be very audible.

An *oscillator* is another general feature on most consoles. It permits levels to be set on the console and on the tape machines, as well as being used in the alignment of the tape machines. In addition, it generates the tones placed at the head of master tapes to ensure equal frequency response from machine to machine or in a tape transfer to disc, cassette, or compact disc.

References

1. Alexandrovich, George, "The Audio Engineer's Handbook," *db, The Sound Engineering Magazine*, Vol. 4, No. 9, pp. 4-6.
2. *Neve 8128/DSP Specification Sheet and Products Bulletin*, Rupert Neve Inc.
3. Sony/MCI, *JH-600 Console Specification Sheet and Operating Manual*, Sony/MCI Corporation.

10 NOISE-REDUCTION DEVICES

The dynamic range of music approaches 120 dB while the dynamic range of a magnetic recording has been limited, until recently, by a signal-to-noise ratio of 60 to 65 dB. The limitations imposed upon conventional analog recorders are due to distortion that is caused by tape saturation (when the recorded level is too high) and tape noise (which intrudes when the recorded level is too low) as illustrated in Fig. 10-1. A desire to reduce the noise level heard in playback causes engineers to either record at high levels (with some distortion) or change the dynamic range of the signal. Both methods are acceptable, with the former method gaining popularity due to low-noise tape formulations.

Fig. 10-1. The dynamic range of recording tape.

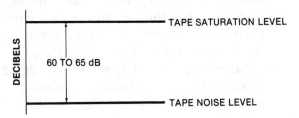

The type of noise that has to be eliminated can be classified as:

- tape and amplifier hiss,
- crosstalk between tracks,
- print-through,
- modulation noise.

Modulation noise is a high-frequency noise that can cause waveform fuzziness as well as the generation of sideband frequencies. This is due to irregularities in the coating of the magnetic recording tape (Fig. 10-2). It is present only when a signal is present and it increases in level as the signal level increases. This modulation or *asperity noise* is a major drawback in the medium of analog magnetic recording.

Fig. 10-2. Modulation noise on a sine wave.

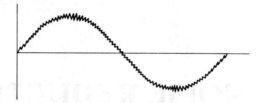

There are several obvious methods which can be used to eliminate these noises:

1. Use amplifiers that have very little hum, buzz, and hiss.
2. Design the tape heads and electronics so that there is very little crosstalk between channels.
3. Improve the noise quality of professional recording tape and/or increase the tape speed so that higher flux levels can be recorded.
4. Make a medium that plays back with no hiss, modulation noise, or print-through.

At present, using the first two methods can bring noise levels down to a point where it is not a limiting factor in the signal-to-noise (s/n) ratio. The third method has occurred in the last 5 years (in analog recording) with the advent of low-noise tapes which improve the s/n ratio by 3 dB. The acceptance of 30 inches per second (ips) as a standard tape speed has lowered this figure even further. The last problem has been virtually eliminated with the arrival of digital recording which reduces hiss, modulation noise, print-through, and self-erasure to values that are below measurable levels.

Although the digital medium is now fully on the horizon, the expense of digital multitrack recording and editing facilities will keep the field of analog recording in our midst for the immediate future. While waiting for price breakthroughs, several methods have been developed to reduce those noises inherent in the analog recording medium; these include: the Dolby system, the DBX system, the single-ended noise-reduction system, and the noise gate.

The Dolby System

Two different Dolby systems are available. The A-type system is used professionally and provides 10 dB of noise reduction below 5 kHz, increasing gradually to a maximum of 15 dB at 15 kHz. The "A" system will reduce any noise that is induced into the signal between the record processor output and the playback processor input. The B-type system is designed for consumer use and it only reduces hiss; it has no effect on low-frequency noises, such as hum, rumble, and pops. The noise reduction is 3 dB at 600 Hz and it rises to 10 dB at 5 kHz, where it levels off.

Dolby systems make use of the same concept as does the NAB pre- and post-equalization curves that are used in analog tape recording. The function of the NAB EQ is to boost the high frequencies before they are recorded (Fig. 10-3A), so that when they are played back, their level can be reduced. Since the high-frequency tape hiss was not boosted by the recording EQ (because it is generated by the tape itself), the playback high-frequency cut decreases the level of the tape noise as it brings the level of the high frequencies back to normal; thus, the high-frequency tape noise is reduced (10 dB in Fig. 10-3B).

Fig. 10-3. Noise reduction by pre- and post-equalization, using 10 dB of high-frequency EQ.

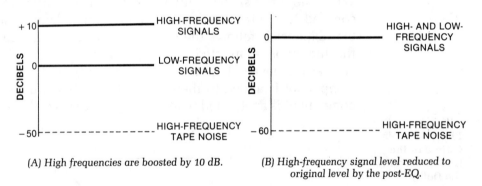

(A) High frequencies are boosted by 10 dB.

(B) High-frequency signal level reduced to original level by the post-EQ.

The pre- and post-EQ of the tape recorders are set to one standard curve (NAB or CCIR) and, because the EQ is fixed, it must enable the recorder to handle the worst-case conditions present in the recording. This means that the tape must not be overloaded by the pre-emphasis when recording any sound source, such as the extreme lows of an organ or the extreme highs of a crashing cymbal. Because saturation must be avoided in these worst-case conditions, the EQ cannot provide optimum boost for each type of program material. For example, an NAB recording of a piano or a viola does not make full use of tape storage capacity in the high-frequency range because the higher harmonics of the piano and viola are much softer than their fundamentals, and most of their output is in the midrange tones. Thus, the high-frequency pre-EQ could be boosted more than the NAB standard without overloading the tape. To play the tape properly, the post-EQ would have to cut the signal by the same amount that the pre-EQ boosted it. This is, in effect, what the Dolby system does. It provides an automatic means of making full use of the tape storage capacity at all times, based on the level of the signal fed to the Dolby unit.

Variable tape loading is achieved through compression in the recording process and expansion in the playback mode. Compression is achieved by increasing the gain of low-level signals rather than by decreasing the gain of the loud signals. The circuit uses a limiter which prevents the signal at its output from rising much above − 40 dBm (Fig 10-4). The output of this limiter is added algebraically to the uncompressed input signal (Fig. 10-5). Since the output of this limiter only rises slightly above − 40 dBm, the effect of the addition of this signal to the input signal depends upon the input signal level. When the input signal is low, the output of the limiter is large in comparison, and adding the two results in a boosted signal. When the input signal is high, the limiter output is small in comparison and its contribution, when the two signals are added, is negligible. Expansion follows the same procedure except that the output of the limiter is subtracted from the input signal, which reduces the gain at low levels (Fig. 10-6). What occurs, in effect, is that an extra component is added to the signal when it is recorded, and the same component is subtracted when the signal is played back.

Fig. 10-4. Output of the limiter used in the Dolby system rises slightly above − 40 dBm.

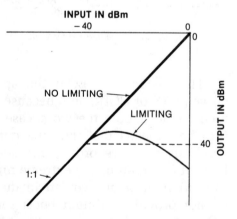

If the signal is processed exactly as just described, full noise reduction would occur only at low signal levels while, at high levels, the noise would have its normal value. In addition, because the level of the noise would be changing, a swishing or breathing sound

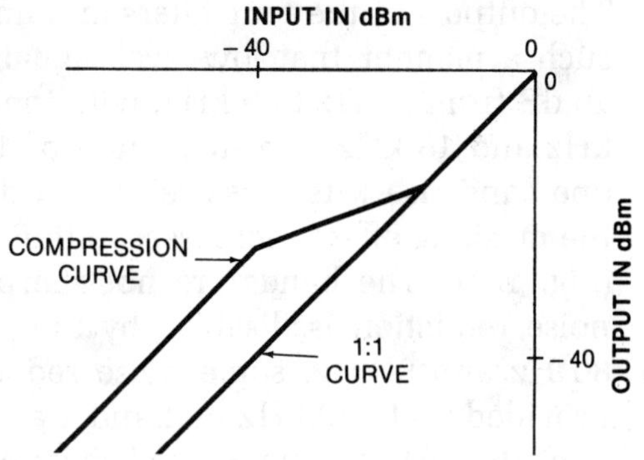

Fig. 10-5. The Dolby compression curve.

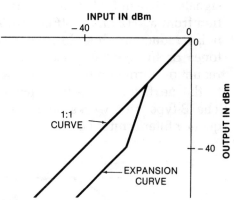

Fig. 10-6. The Dolby expansion curve.

would be heard. The Dolby system takes advantage of the masking effect of the human ear to avoid these unacceptable results. Since the ear cannot hear noise in the same frequency range as the signal, only noise frequencies outside the signal bandwidth are responsible for the swishing and noise heard during loud passages.

The A-Type System

In order to reduce these noise frequencies, the audio spectrum is divided into four bands in the A-type system. Each band of frequencies has its own limiter so that the presence of a loud signal in one band does not defeat the noise reduction in the other bands. The four bands (Fig. 10-7) are:

1. 80-Hz low pass (80 Hz and below).
2. 80-Hz to 3-kHz bandpass.
3. 3-kHz high pass (3 kHz and above).
4. 9-kHz high pass.

The outputs of the four filters and limiters (Fig. 10-8) are combined in such a manner that low-level signals (below − 40 dBm) are boosted 10 dB from 20 Hz to 5 kHz, with the boost rising gradually between 5 kHz and 15 kHz to a maximum of 15 dB. As the level of a signal in one band rises, its noise reduction decreases, but the effectiveness of the masking effect increases, so the noise level appears to be constant (Fig. 10-9). The bands are not sharply defined, so when the Band 2 noise reduction is disabled by the presence of loud signals between 80 Hz and 3 kHz, some noise reduction (in addition to masking) is provided up to 120 Hz by Band 1 and, also, above 1.8 kHz, by Band 3 (Fig. 10-10A). If Band 3 also has its noise reduction turned off by loud signals between 3 kHz and 9 kHz, Band 4 will contribute noise reduction from 5 kHz and up (Fig. 10-10B). Bands 1 and 4 rarely have their noise reduction shut down completely, except by very loud organ tones or by cymbal crashes. Although the actual amount of noise reduction throughout the audio spectrum changes from one moment to the next, the noise level perceived by the ear remains constant. The B-type unit works similarly but it uses only a single high-frequency filter/limiter band.

Fig. 10-7. The four filter bands of the A-type Dolby system.

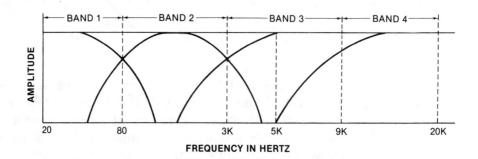

The operation of the Dolby system depends upon its being used both when a tape is recorded and when it is played back. Each Dolby channel has a *record* and a *play mode,* and the processing can be inhibited through the use of the *noise reduction in/out switch.* When this switch is in the "out" position, the Dolby unit becomes a unity gain amplifier. The modes are switched by controlling the use of the combined signal output of the filter/limiters. This output is added to the input signal in the record mode, subtracted from the input signal in the play mode, and is not used when the noise-reduction switch is in the "out" position.

Each track of a tape is assigned a separate Dolby unit. The same unit that processes the signal for recording is used for playback pro-

Fig. 10-8. Block diagram of the filter/limiter networks.

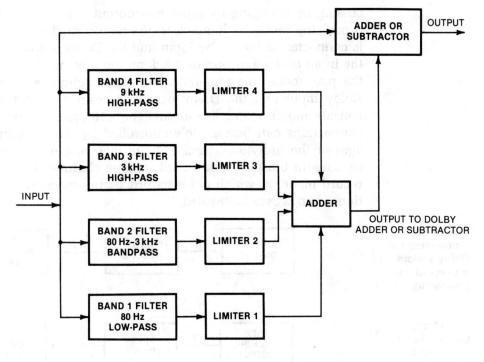

Fig. 10-9. The masking effect and the actual noise reduction combine to keep the apparent noise level constant.

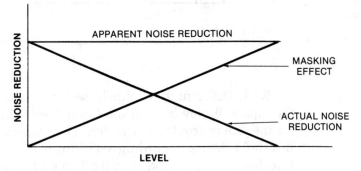

Fig. 10-10. Demonstrating noise reduction in the A-type system.

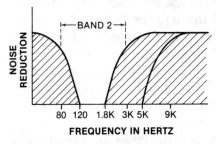

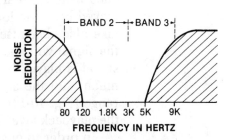

(A) Band 2 noise reduction is shut off by a high-level signal between 80 Hz and 3 kHz.

(B) High-level signals between 80 Hz and 9 kHz shut off the noise reduction in both Bands 2 and 3.

cessing by changing its input and output connections whenever the processing mode is changed. In the record mode, the console output is connected to the Dolby input and the Dolby output is connected to the input of the appropriate track on the tape machine (Fig. 10-11). In the play mode, the tape-machine signal output is connected to the Dolby input and the Dolby output is connected to the appropriate console monitor and line inputs (Fig. 10-12). The mode and signal connections can be remotely controlled by the recording relay voltages of the individual tracks of the tape machine. Each Dolby unit remains in the play mode until its tape channel is punched into the record mode, at which point its processing mode and signal connections change over to "record."

Fig. 10-11. Connecting the Dolby system for record processing.

Fig. 10-12. Connecting the Dolby system for playback processing.

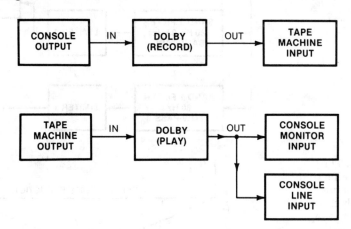

Each Dolby unit is initially set up so that a 0-VU level signal on the tape will play back at the *Dolby level* indicated by the NAB mark on the Dolby front-panel meter. This ensures that signals below the threshold during recording will also be the same amount below the threshold on playback. A ±3-dB tolerance will exist before a difference in the signal playback becomes very noticeable. If the signal is played back too loud through the Dolby unit, too little expansion will take place because the limiter is blocked by the high-level signal and the signal will sound compressed and overly bright (Fig. 10-13). If the signal is played back at too soft a level, the expander will expand too much, and the signal will sound dull and have too great a dynamic range (Fig. 10-14). The signal record level does not matter as long as the playback level is the same.

In order to ensure that the tapes are played back at the same level at which they were recorded, a 400- or 700-Hz, 0-VU level tone is recorded onto the beginning of the tape so that the reproduction level

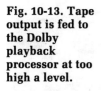

Fig. 10-13. Tape output is fed to the Dolby playback processor at too high a level.

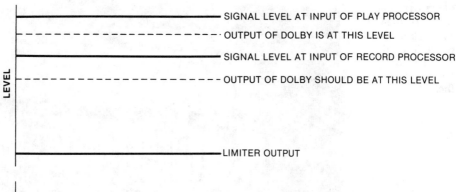

LEVEL

——————————————— SIGNAL LEVEL AT INPUT OF PLAY PROCESSOR

– – – – – – – – – – – – – – OUTPUT OF DOLBY IS AT THIS LEVEL

——————————————— SIGNAL LEVEL AT INPUT OF RECORD PROCESSOR

– – – – – – – – – – – – – – OUTPUT OF DOLBY SHOULD BE AT THIS LEVEL

——————————————— LIMITER OUTPUT

Fig. 10-14. Tape output is fed to the Dolby playback processor at too low a level.

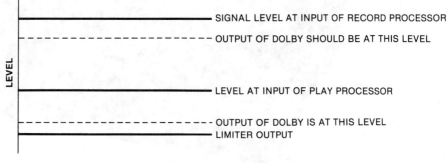

LEVEL

——————————————— SIGNAL LEVEL AT INPUT OF RECORD PROCESSOR

– – – – – – – – – – – – – – OUTPUT OF DOLBY SHOULD BE AT THIS LEVEL

——————————————— LEVEL AT INPUT OF PLAY PROCESSOR

– – – – – – – – – – – – – – OUTPUT OF DOLBY IS AT THIS LEVEL

——————————————— LIMITER OUTPUT

can be adjusted properly when the tape is played back. This is especially important if the tape is to be interchangeable with those made at other studios where the machine playback levels might have been set differently. Dolby units have built-in oscillators which emit a reference tone, at the Dolby level, that can be recorded on tape. The oscillators have a pulsating characteristic which identifies the tape as being Dolby processed.

The Dolby system (Fig. 10-15) can be used in any audio storage or transmission chain if the signal is available for processing at both ends. It can be used for discs, telephone lines, radio, etc. The processed recording signal is referred to as the *stretched* or *encoded* signal. The playback signal is called the *unstretched* or *decoded* signal. An encoded signal is not distorted. Its waveform is the same as the original signal; only its amplitude is changed. If it is stretched, the signal sounds very bright and compressed, but it is still intelligible. When making a stretched copy of a stretched tape, no processors are necessary; the signal retains its noise-reduced quality and the copy can be made directly from machine to machine. The same amount of noise increase occurs in making tape copies of a stretched tape whether it is decoded and, then, re-encoded (Fig. 10-16A), or transferred direct from machine to machine without processing (Fig. 10-16B).

If it is desired to equalize a stretched signal, or mix a stretched signal with another stretched signal or with an unstretched signal, all the signals must first be restored to an unstretched state. If the stretched signals are compressed, equalized, or changed in level through mixing, the signals may be changed to the point that they cannot be unstretched properly.

The B-Type System

Dolby makes only one B-type processor, the Model 320 (Fig. 10-17). It is designed for use in making master tapes which will be used in the high-speed duplication of cassettes and reel-to-reel tapes. All other B-type processors are made by manufacturers under licence from Dolby Laboratories, Inc.

The DBX System

The DBX noise reduction system (Fig. 10-18) is a compressor/expander system which provides about twice as much noise reduction as the Dolby system. Noise is reduced to 20 or 30 dB below what it would otherwise be. The system can also be considered as a means of loading the tape more effectively than is possible with the NAB curves. It is connected between the console and the tape machine in

Fig. 10-16. The same amount of noise increase occurs in making tape copies of a stretched tape.

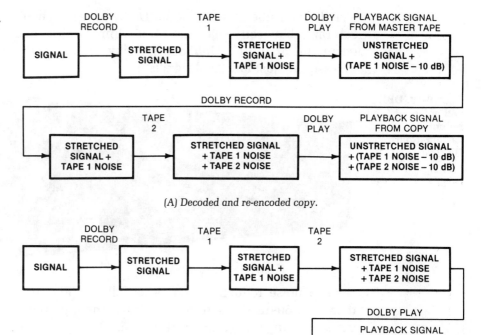

(A) Decoded and re-encoded copy.

(B) Direct transfer copy.

Fig. 10-17. The Dolby Model 320 B-type processor *(Courtesy Dolby Laboratories, Inc.).*

exactly the same manner as the Dolby system. The compressor has a 2:1 ratio between −90 and +25 dBm, with the unity gain point occurring at 0 VU. Unlike the Dolby system, all signals are compressed, not just low-level signals.

Fig. 10-18. DBX 208 noise-reduction system *(Courtesy DBX Inc.)*.

The noise reduction occurs as follows (Fig. 10-19): Assume that there is a 60-dB signal-to-noise ratio in the tape recorder and a 60-dB dynamic range program is recorded. During the softest passages of the program, the noise added by the program is just as loud as the program and is, therefore, very audible. With the DBX system, the program passes through the DBX record section and is compressed 2:1 into a 30-dB dynamic range program (Fig. 10-19A). This compressed signal is then recorded onto tape where it is, in effect, mixed with the tape noise. The tape noise is now, however, 30 dB below the softest passage of the music (Fig. 10-19B). On playback, the expander reduces the signal level 30 dB on the softest passages and reduces the noise 30 dB at the same time, so the signal is now 60 dB below 0 VU and the noise is 30 dB below that, or 90 dB below 0 VU (Fig. 10-19C). Thus, the noise reduction is 30 dB.

The difference in level between the noise due to the tape and the softest signal recorded on the tape determines the signal-to-noise ratio. In order to achieve more than 30 dB of noise reduction with the same program, the compression ratio would have to be increased. For example, a 3:1 ratio would compress the signal down to a 20-dB dynamic range so that it would be 40 dB above the noise level, and 40 dB of noise reduction would result after expanding the signal. A 2:1 ratio was chosen over the 3:1 ratio because of the effect of tape drop-outs on the expander. Since the expander reduces its output signal twice as much as the signal played back from the tape, a tape dropout of 2 dB causes a 4-dB dropout of signal with the 2:1 ratio. A 3:1 ratio would cause a 6-dB dropout. A ratio of 1.5:1 would make the dropout

Fig. 10-19. Noise reduction in a DBX system.

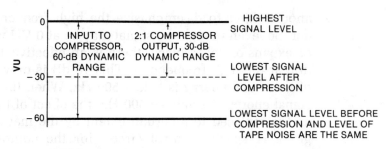

(A) A 60-dB dynamic range compressed to 30 dB before recording.

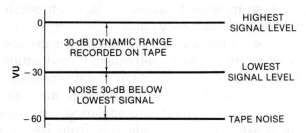

(B) When recorded, softest part of program is 30 dB louder than tape noise.

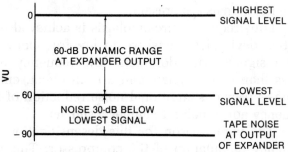

(C) Dynamic range is expanded on playback. Program signal is reduced from −30 to −60 dB, noise is reduced from −60 to −90 dB.

problem less noticeable but it would do so at the cost of noise reduction (only 20 dB possible). The 2:1 ratio was considered to be the best compromise between the amount of noise reduction and the sensitivity to dropouts.

The next difference in the DBX system is that it operates on the full frequency range from 20 Hz to 20 kHz. Dolby Laboratories found that having the compressor operate over the full frequency range caused a loss of noise reduction throughout the frequency spectrum in the presence of loud signals, even though they might have a restricted frequency range. The noise not masked by the signal became apparent on loud passages. The DBX system overcomes this problem to a certain extent by using a filter to pre-emphasize the high frequencies by 12 dB before the compressor and by using another filter to de-emphasize the high frequencies after the expander. So, even though the signal may be at 0 VU when no compression or expansion takes place, there is an effective 10 dB of noise reduction at the high frequencies. This 10 dB is only achieved when most of the signal energy is below 500 Hz. When the major portion of the signal energy rises above 500 Hz, the effect of the filter is diminished and is replaced to a certain extent by the masking effect of the program. A special control circuit for the compressor and expander reduces the gain when the level of high-frequency signals becomes very high in order to avoid the tape saturation that would be caused by NAB high-frequency boost.

An even more important reason for the use of the high-frequency pre-emphasis is the high-frequency modulation noise of tape, which increases in level as the signal level increases. At 0 VU, the modulation noise would be very noticeable and, since the expander constantly changes gain, the modulation noise also changes noticeably. The pre-emphasis prevents this from becoming obtrusive, but the variation is noticeable at very high listening levels when only low-frequency signals are present.

Low-frequency pre-emphasis is achieved by making the level-sensitive device less sensitive to low-frequency signal components, so that signals with dominant low-frequency energy are compressed less than other signals and are, in effect, emphasized. This pre-emphasis provides additional noise reduction of hum and other low-frequency noises behind high-level signals.

The main reason for this desensitization is to eliminate low-frequency modulation of the compressor. This could occur if frequencies in the 3- to 5-Hz range (due to subways, trucks, and air conditioning, which can cause buildings to vibrate) are picked up (by vibration) through the mike stands and transmitted to condenser mikes, which have a good low-frequency output. If the level sensor was not desensitized to these frequencies, they would trigger the compressor. However, since most professional tape recorders will not pass such low frequencies, they would not be present at the input to the expander to control its gain. Consequently, the background noise would rise and fall at a 3- to 5-Hz rate. The desensitizing amounts to 3 dB at 50 Hz and 20 dB at 5 Hz.

The DBX level sensor computes the true rms value of the input signal and uses this value to determine how much compression or expansion should be applied. The Dolby system uses a combination of *peak values* and a "type of" *averaging values* (different from rms) to approximate rms sensing. Peak and non-rms values can be changed by the phase shifts and frequency-response errors that occur in amplifiers and tape machines. Rms values are also changed by these defects but by a much smaller amount, so that the DBX compressed signal does not loose its control information when passed through amplifiers and tape heads. The Dolby-stretched signal-level control information can be changed by these devices so that the "unstretching" may not exactly remove the "stretching." As with the Dolby system, DBX tapes can be copied in the encoded state and the copies can be played back through a decoder.

Single-Ended Noise Reduction

The systems discussed up to this point all require processing in both recording and playback to achieve noise reduction. Noise cannot be removed from a signal with these devices; it can only be prevented from entering. Recently, there have been a number of entries into the noncomplementary or single-ended noise-reduction market. Single-end noise reduction acts basically as a frequency-dependent expander or *noise gate*. The Symetrix Model 511 (Fig. 10-20) works to delete noise from audio sources by utilizing a downward dynamic range expander in conjunction with a program-controlled dynamic low-pass filter. The downward expander and dynamic filter may be used together or separately to provide the greatest possible amount of noise reduction.

Fig. 10-20. Symetrix Model 511 single-ended noise reduction unit *(Courtesy Symetrix Inc.).*

The dynamic expander consists of a voltage-controlled amplifier (vca) and associated detection circuitry. The expander reduces broadband noise by attenuating the signal whenever the level falls below the threshold point (adjustable on the front panel); this results in an increased dynamic range over the entire audio spectrum. The downward filter of the Model 511 accomplishes noise reduction by taking advantage of several basic psychoacoustical principles. The first principle is that music is capable of masking noise within the same bandwidth. The second principle is that reducing the bandwidth of an audio signal reduces the perceived noise because the greater the spectral distribution of noise, the greater is the human ear's sensitivity to that noise. The dynamic filter examines the incoming signal for high-frequency content and, in the absence of high-frequency energy, the filter bandwidth decreases to as low as 2 kHz. When high-frequency energy returns, the filter opens back up as far as is necessary to pass the entire signal.

Noise Gates

As we have seen from the section on signal-processing equipment, a *noise gate* may often be a very effective noise-reduction device when used in reducing background noise on certain program material. Operating on the masking principle, a noise gate may be viewed as a "high-ratio" expander, which acts as a unity gain amplifier in the presence of a signal above a certain set threshold, passing the signal at full volume. When the audio signal at the input of the device falls below the preset threshold level, the signal is effectively turned off at the output. Thus, noise is eliminated from the signal at the point where there is no other signal present to mask it. For critical program material, it may be necessary to "fine tune" the *attack* and *release* controls to eliminate unwanted "pumping" or "breathing" of the noise floor below the desired signal.

References

1. Blackmer, David E., "A Wide Dynamic Range Noise-Reduction System," *db, The Sound Engineering Magazine*, Vol. 6, No. 8, August/September 1972, pp. 54-56.
2. *DBX 187 Technical Bulletin*, DBX, Inc.
3. *Dolby A301 Operating Manual*, Dolby Laboratories, Inc.
4. *Symetrix 511 Technical Bulletin*, Symetrix Inc.

11 INTERLOCKING TAPE MACHINES

Over the last decade, the recording industry has changed a great deal; one of the major transformations has been the incorporation of the audio, video, film, and electronic music media. Motion pictures have been an integral part of the recording industry for many decades and, within the last five years, we have seen the strong emergence of video sound and its incorporation into the world of the multitrack recording studio. Also, we have seen the beginnings of the electronic instrument boom, incorporating the Musical Instrument Digital Interface (MIDI) for use in musical instrument synchronization. In order for these multiple mediums to work together simultaneously, it is of extreme importance that the many tape transports and other equipments used be locked together in time. The method which allows multiple audio and visual media to maintain a direct relationship to one another, relative to time, is known as *synchronization* (Fig. 11-1).

Synchronization

Synchronization, or sync as it has come to be known, is achieved by *interlocking* the speed or clocking rate of two or more machines, allowing them to run together in simultaneous operation. Sync does not require that all the machine speeds or clocking rates be constant, but it does require that they be equal at all times. If one machine slows down or speeds up relative to the other, synchronization is lost, but if both machines slow down or speed up by the same amount simultaneously, sync is maintained. Neither hysteresis synchronous motors nor servo motors, alone, can maintain tape speed accurately enough for this application. Both of these speed-control methods maintain a constant capstan rotational speed and rely on the pressure of the pinch roller or the pressure on the holdback tension arms to transmit this into a constant tape speed. Since varying

**Fig. 11-1.
Diagram of
interconnected
transport
machines.**

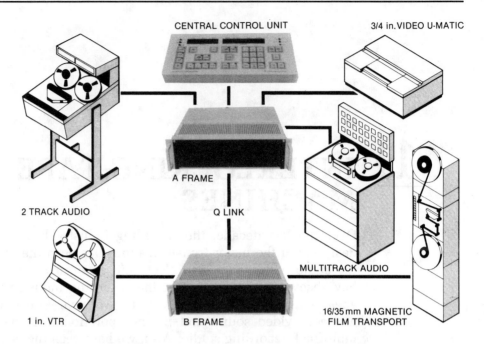

CENTRAL CONTROL UNIT

3/4 in. VIDEO U-MATIC

A FRAME

2 TRACK AUDIO

Q LINK

MULTITRACK AUDIO

1 in. VTR

B FRAME

16/35 mm MAGNETIC
FILM TRANSPORT

amounts of slippage between the tape, capstan, and the tension arms can occur with this method of moving tape, fluctuations can be introduced into the tape speed.

In order to keep equipment in sync, a means of measuring and correcting the relative speeds involved must be used. This is accomplished by recording a *sync tone* on one track of each tape; this is called the *sync track* or *control track*. Any variation in the speed of the tape during record or playback of this tone will cause a change in sync tone frequency. Through the use of a *resolver*, the control-track signal frequency of one machine can be compared with that of an *external sync reference*, such as the sync tone generator, and its capstan motor speed can be varied to make the frequencies equal. The tape speed is controlled in such a way that each cycle of the control-track signal is brought precisely into phase with each cycle of the external signal. This is called *phase-locked* operation. If the external reference frequency is the same as that originally recorded on the control track, the information on the tape will play back at exactly the same tempo at which it was performed. The sync tone is often thought of as providing *magnetic sprocket holes* on the tape, which are driven cycle by cycle by the external sync reference in the same manner that camera sprockets drive the holes in motion-picture film.

For use with a hysteresis synchronous motor, the resolver produces a 60-Hz output signal which feeds a power amplifier, which, in turn, drives the capstan motor. If the control-track frequency is lower than the external sync frequency, the resolver output frequency increases to speed up the capstan. If the control frequency is higher than the external reference, the output frequency decreases to slow down the capstan.

Servo motors have resolvers built into them and constantly compare the output of their tach head with a reference frequency. If the sync track, instead of the tach head signal, is fed to the servo motor, the motor speed varies to keep the tape speed, rather than the capstan speed, constant. Although machines using different sync frequencies can be synchronized by using a separate resolver and external reference frequency for each machine, it is more common in audio work for all machines to use the same sync frequency. The most commonly used sync tone frequency is 60 Hz since it is available wherever a power line is accessible.

The standard system for interlocking audio, film, and video transports is the SMPTE (Society of Motion Picture and Television Engineers) serial time code. This code, designed for use in electronically editing video tape, uses an 80-bit digital word which is recorded at the beginning of each video frame. Each word assigns a different address to the corresponding point on the tape, expressed as the number of elapsed hours, minutes, seconds, and frames from the beginning of the program. Since there are 30 black-and-white video frames per second, new words are generated 30 times a second, producing a bit rate of 2400 bits per second. The code can be adapted to the color video standard of 29.97 frames per second by having the synchronizer count every thousandth frame twice (called the "dropped-frame technique").

The SMPTE time code uses the digital phase-modulation format called *biphase mark*. Each digital word is divided into time intervals called *bit cells,* and one bit is stored in each cell. A "1" is represented by a magnetic flux polarity reversal at the middle of the cell, while a "0" is represented by the lack of a reversal at this point. All cells have flux reversals at their beginning (or end).

In Fig. 11-2, the word shown corresponds to a readout of 16:47:31:23 on a time code reader. Bit groups A, B, C, D, E, F, G, and H are for optional binary control words which can be stored within the code to control auxiliary functions. The bits are left as 0's when no optional words are used.

The last 16 bits form a *sync word* which is used in establishing the start of the next frame, as well as in adjusting the bit counting rate so that the address can be read when the tape is run at higher or lower

Fig. 11-2. The 80-bit time code word consisting of binary ones and zeros.

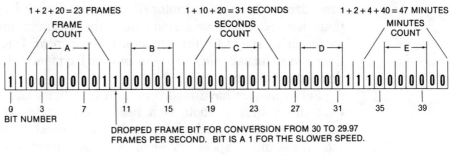

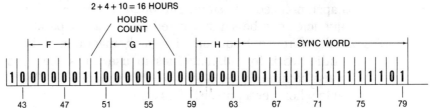

than normal speed, or in reverse. A SMPTE *time code generator* is used to record the code onto the control tracks of the machine transports involved.

A frame-by-frame lock can be achieved between two or more machines by the use of a *time code synchronizer* (Fig. 11-3). Synchronization is achieved by recording of time code onto the control tracks of all the machines to be interlocked. This may either be recorded beforehand, generated at the time of recording, or generated when dubbing from one machine to another. Once this is accomplished, the synchronizer is assigned a master recorder and one or more slave recorders. The synchronizing process then will read the SMPTE time code address from the master recorder and compare this address to the relative addresses of each slave recorder (Fig. 11-4). A dc offset voltage is then fed to the slave servo speed mechanisms, changing the relative slave speeds until all addresses, which are read at the synchronizer, read the same time code value. At this point, synchronization is achieved and the slave recorders will follow the master recorder in precise sync under all conditions. If the slave is behind the master, the synchronizer senses a lower address on the slave tape and speeds it up until it matches the master address. If the master is behind the slave, the slave is slowed down until the master catches up. For sync to be achieved, the codes on the control tracks of the master and slave tapes must be identical and, with certain older synchronizer models, the relative positions may need to be within the synchronizer *capture range*.

Fig. 11-3. MCI JH-45 Autolock synchronizer *(Courtesy Sony/ MCI Corp.).*

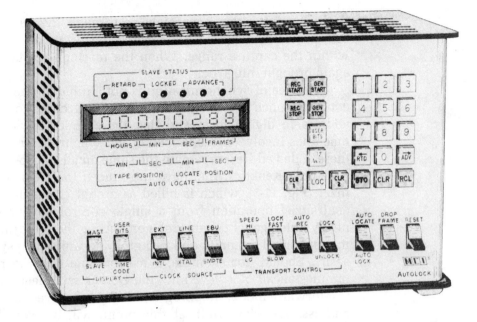

Fig. 11-4. Synchronizing two tape machines using the SMPTE code.

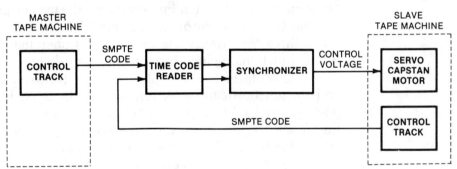

A *time code reader*, which can be switched to read either the master or slave code, produces a numerical readout of the address in hours, minutes, seconds, and frames to enable the engineer or tape operator to *park* the master and slave tapes within the capture range of the synchronizer. The operator places the master tape in the play mode, and the synchronizer automatically starts the slave from the park position upon sensing the presence of the master code.

The tapes are brought into absolute sync and held there regardless of any tape stretching which may have occurred after the control tracks were recorded. When the master is stopped, the synchronizer stops the slave machine. Since the slave is stopped within the capture range, sync is quickly established when the master is started again. If the tapes are to be played again, they are rewound to the desired point, either separately or through the autolocate functions that are available on newer synchronizer models, and are parked within the capture range. When the master is restarted, the slave is again brought into sync.

The idea of synchronization is not a difficult one to understand in actuality, once the idea that the SMPTE code contains a location address in digital form and that these codes are recorded onto all machines involved. It is the job of the synchronizer to make sure that these coded addresses remain the same on all transports at all times.

Audio Kenetics, Inc. have introduced their Q-Lock 3.10 synchronizer (Fig. 11-5), which is billed as a control synchronizer. It allows centralized operation from a single control keyboard. This unit is highly intelligent, capable of comparing the sync addresses of the master and slave units and bringing the units into sync quickly and easily. Should, for example, a master time code be read as being 3 minutes ahead of a slave unit, the Q-Lock will place the slave into the fast-forward mode, stopping it within a few seconds of the master address. The slave unit, at this point, will be placed back into the "play" mode and brought into sync with the master unit.

The Q-Lock synchronizer permits complete remote control over up to three tape transports and may be operated as an autolocator whose functions include instant replay, cycle (continuous replay of a section), locate to cue and play, etc. Control over the recording in-and-out function can be by remote control as each machine's transport record facilities may be accessed either manually or by using time code address entry and exit points. A record advance feature allows for the accurate rehearsal of record drop in/drop out points before committing to an actual take.

The Q-Lock system may be used to synchronize up to three machines, with an option of extending this to five transports by the use of an optional Q-Link "add on." The designation of the system master is easily reassignable from the central keyboard. The time code generator and reader is capable of working with a multitude of frame-per-second standards, including drop-frame. *Time code offset* may be introduced in order to advance or lag the sync by a specific amount of time, with the numeric keyboard allowing the easy calculation and entry of offset values. An *offset retention mode* will automatically update offsets when discontinuities, such as edits on encoded tape, occur.

One feature offered by the Q-Lock system is the *Dedicated Machine Interface*. One of the major difficulties involved in the use of a modern synchronizer is the multitude of differing transports that it

Fig. 11-5.
Q-Lock 3.10
synchronizer
(Courtesy Audio
Kenetics, Inc.).

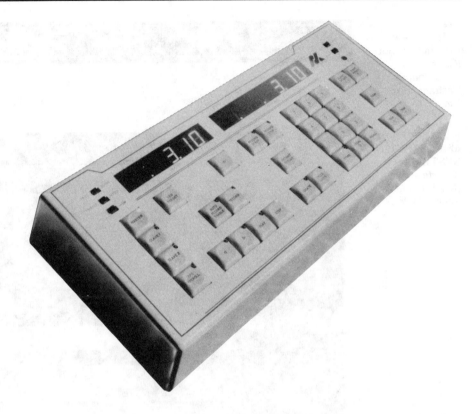

must accommodate, with many transports operating in differing formats or with different principles of transport control. The dedicated machine interface connects the Q-Lock to each type of machine by using an interface which is especially designed for a specific machine transport and logic.

Synchronous Technologies, Inc. have developed their SMPL™-Lock interface panel (Fig. 11-6). Originally designed for the smaller studio, but gaining wider acceptance, it is a computer-based system which integrates a chase-lock synchronizer, a 10-point autolocator for audio- and video-tape transports, an MIDI (Musical Instrument Digital Interface) electronic instrument synchronizer, a time code-derived metronome, and an SMPTE time code generator and reader which works in all film- and video-frame rates, including drop-frame. The SMPL™ Lock system consists of a control console (Fig. 11-7) which displays system information on either a video monitor or a tv set and a separate rack-mounted panel (Fig. 11-6) with interface connectors for the time code, tape transports, and MIDI buses.

The SMPL™ Lock synchronizer is a chase system which requires that the slave transport "chase the master under all conditions."

**Fig. 11-6.
SMPL™Lock
interface panel**
*(Courtesy
Synchronous
Technologies,
Inc.).*

**Fig. 11-7. Photo
of the SMPL
SYSTEM™**
*(Courtesy
Synchronous
Technologies,
Inc.).*

Thus, if the master is 3 minutes ahead of the slave, the slave will speed up until it catches up with the master, at which time, a time sync lock is achieved.

This system also allows for *automatic record punch-in/-out points*, which are preset by the time code address on the tape. A selection of 8 *autolocate points* and a tape-machine remote control is combined with an automatic looping function which will automatically rewind the multitrack tape to the desired autolocate point and then repeat the play or rehearse mode. These autolocate points may also be programmed to be *event outputs*, which are switching functions for triggering effects, changing instrument presets, and setting console channel muting.

The SMPL SYSTEM™ (Fig. 11-8) also offers a MIDI (Musical Instrument Digital Interface) output which can be used to precisely synchronize electronic drum sets, sequencers, and digital keyboard recorders with the material recorded onto tape, thus providing a synchronized output with pulses occurring at the contemporary standard of 24 pulses per metronomic beat. With this feature, it is possible to sync electronic instruments to the SMPTE control track on the multitrack tape in mixdown, thus replacing the audio tracks on the tape with a first-generation synthesized signal direct to the final master tape.

**Fig. 11-8.
Diagram of the
SMPL system.**

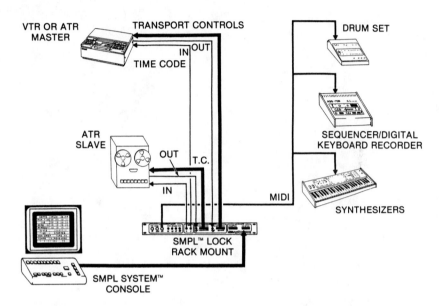

References

1. DeWilde, Carl, "A New Approach to Synchronization and Spliceless Editing," *Audio Engineering Society Preprint No. E-3*, 46th Convention, September 10-13, 1973.
2. King, Marshall, "The Sync Pulse in TV Audio," *db, The Sound Engineering Magazine*, Vol. 5, No. 8, August 1971, pp. 19-23.
3. _____, "The Sync Pulse in TV Audio, Part 2," *ibid.*, Vol. 5, No. 9, September 1971, pp. 25-28.
4. _____, "The Sync Pulse in TV Audio, Part 3," *ibid.*, Vol. 5, No. 10, October 1971, pp. 23-29.
5. Nothaft, Eugene M., and Tom W. Irby, "Multitrack Audio in Video Production," *Journal of the Audio Engineering Society*, Vol. 21, No. 3, April 1973, pp. 172-176.
6. Silver, Sidney L., "Synchronous Recording Techniques," *db, The Sound Engineering Magazine*, Vol. 4, No. 7, July 1970, pp. 25-29.

7. *SMPL Synchronizer Technical Bulletin*, Synchronous Technologies, Inc.
8. *Q-Lock Technical Bulletin and Operating Manual*, Audio Kenetics, Inc.

12 AUTOMATED MIXDOWN

In multitrack recording, the final musical product is not realized until the mixdown stage. Each track of the tape has its own volume, echo send, front-to-rear and left-to-right pan pot, as well as several EQ frequency and amount of boost (or cut) selectors. Multiply these functions by the twenty-four or more tracks of instrumentation currently used in recording, and add the auxiliary and echo return volumes, compression/limiting, and other signal-processing functions, and it becomes obvious that the mixdown process may be more than what one engineer can efficiently handle. As a result, mixdowns must be rehearsed repeatedly so that the engineer can learn which controls must be operated, how much they should be varied, and at what point any changes come, relative to the recorded tracks. It is not uncommon for 12 hours to be spent mixing a complicated piece of music before an acceptable mix is obtained. Mixes must often be rejected because the engineer simply forgot to make one important control change. The producer and engineer know how and when the control settings should be changed, but the memory and physical dexterity required to execute them can exceed human capabilities.

The solution to this problem is the use of equipment that can remember and re-create any settings and changes made by the engineer, while allowing him to improve on them one by one until he achieves the desired final mix. This technique is called *automated* or *computerized mixdown* and is made possible through the use of *voltage-controlled amplifiers (VCAs)*, *voltage-controlled equalizers (VCEs)*, *programming encoder/decoders*, and a data storage device called a *memory*. A VCA controls the level of audio signal as a function of a dc voltage applied to its control input, while a VCE changes the frequency response of an amplifier as a function of the dc voltages applied to its control inputs. In a console fully equipped for automated mixdown, VCAs, VCEs, and digital routing devices perform the function of faders, equalizers, and switches, while the actual fader, equalizer, and switch controls are used to vary the dc control voltages.

Automated Consoles

VCAs permit the *free grouping* of signals. Since there is no leakage of the audio signal into the VCA control input, or vice versa, the dc voltage at the control input of any number of VCAs can be controlled by a single variable-voltage source while maintaining complete separation of the audio signals being controlled. The advantage to this is that after the levels of several instruments are set relative to each other, their overall level can be controlled by a single fader rather than having to change each individual instrument. Some methods of this technique are illustrated in Fig. 12-1. In Fig. 12-1C, the separation is retained by using VCAs and a grouping fader to control the level of the voltage fed to the individual dc control faders which, in turn, feed the VCA control inputs. In Fig.12-1D, the output of each VCA is proportional to the sum of its channel and the group control signals. Attenuation is 0 dB when the sum of the control voltages is 0 volts. As the sum increases, so does the attenuation.

In a console not using VCAs, this can be done in two ways: either the signals are mixed into one channel of the console and the mixture is fed through a fader to control the overall level, or the signals are kept separate and are patched into a multichannel fader. In the latter method, the number of tracks controllable by one fader depends on the number of channels available in the multichannel fader. With voltage control, the levels of three instruments placed into a single audio channel may be controlled by a single-channel *submix* or *grouping* fader, also known as a *subgroup*. This is done by controlling the dc voltage feeding the grouping fader; the VCAs can all be switched on or off simultaneously with one switch. For example, if a horn section recorded on several different tracks is to begin playing in the middle of a song, but a few of the horns come in slightly early, the engineer could leave their VCAs at full attenuation until the moment the horns are to play and, then, could turn them all on, simultaneously, with one master switch. Alternately, if one track had no noise and began to play at the proper time, a noise gate could be connected to it and the noise-gate control voltage could be used to turn on all the other tracks when the noiseless track begins to play. Control inputs are grouped by providing several dc grouping buses on the console. All VCAs assigned to the same bus can be controlled by a single voltage source.

Another method of accomplishing automation is with the Neve Computer Assisted Mixing System (NECAM). Found exclusively in Neve consoles, this system operates by way of a touch-sensitive servo-motor-driven fader. Unlike a VCA-based system where the dc voltage levels control the automated operation, the NECAM fader is a resistive attenuator which accomplishes automation by way of an interface that is servo motor driven; the interface moves the fader by a desired amount to the desired position. Thus, during a replay of an automated mix, the faders will move automatically by themselves.

Billed as an instinctive automation system, the NECAM 96 automation system does not have separate provisions for reading, writing, or updating the level of information. Instead, all that is required, once the desired rough mix is achieved, is for the operator to touch the fader to be updated and move it to the desired location. The processing computer will then automatically store this information as updated material.

Grouping of faders, in any number of groups, is assignable. This allows groups to be moved in unison steps simply by touching any fader in that group. Muting, mute groups, and up to 128 event-switching functions may be called up by the control keyboard and stored in the processor. Automated data may be stored as a virtual mix (allowing for an unlimited amount of updates to be stored in a temporary memory), stored on a disc drive (making a permanent record), or stored as one of the snapshots (up to 999) of fader and mute functions for instant call-up.

Modes of Operation

Write Mode

The automated console has basically three different modes of operation: write, read, and update. In the *write mode*, the programmer scans the dc control input level of each of the voltage-controlled devices (VCAs, VCEs, and switching functions) of the console in a certain sequence many times per second and *encodes* this control voltage information into a form that can be stored for later use. The memory unit used for storage is usually an audio tape recorder, and the track on which the information is recorded is called the *data track*. Since each control voltage represents the position of a console control, the data track contains a record of how the console controls were set during each scan cycle. In order for the settings to be reconstructed, the data must be synchronized with the program material, and, therefore, it is recorded on either a spare track of the multitrack master or on a separate tape machine that is interlocked with the master.

Read Mode

In the *read mode*, the console controls no longer supply dc voltages to their respective control inputs. Instead the programmer reads the encoded control information from the data track and feeds it through its *decoder* section which reconverts it into dc voltages and connects these voltages to the appropriate voltage-controlled device inputs. As the master tape and data track are played, the dc voltages control the console level settings producing a mix identical to the one that originally produced the data track. If, after listening to the tracks mixed automatically in the read mode, the engineer and producer decide certain functions need to be completely changed, the write mode can be initiated on these functions only, while retaining automatic control of the others by leaving them in the read mode. The dc voltages created by the decoder and by manual operation of the console are

Fig. 12-1. Varying the levels of several signals with one control.

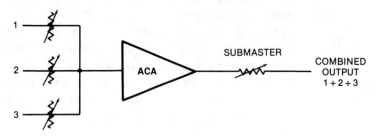

(A) By mixing them together and using a submaster.

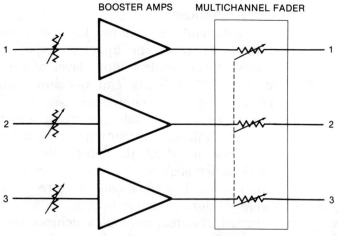

(B) Retaining separation by using a multichannel fader.

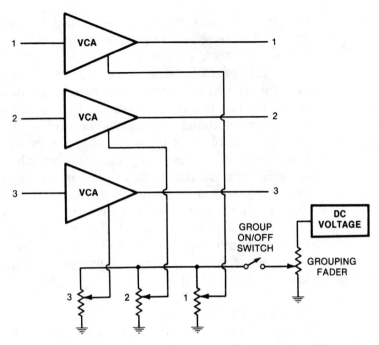

(C) *Retaining separation by using VCAs and a grouping fader.*

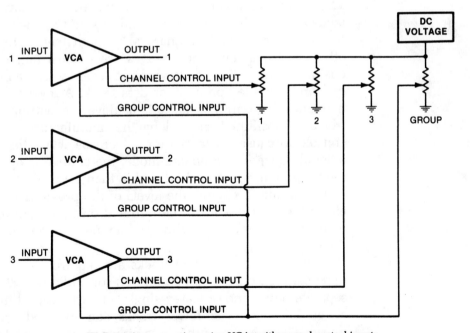

(D) *Retaining separation using VCAs with several control inputs.*

then encoded and stored on a second data track. By alternating back and forth between two data tracks, improvements can be made until the desired mix is achieved.

If only part of the mix needs to be reprogrammed for a certain function, the write mode can be entered on that function at the point where the changes are to be made. In order to provide a smooth transition from automatic to manual control with no jumps in level, the manual control must be set so that the dc voltage it creates equals that which is re-created by the decoder. This is done by manually adjusting the control with the aid of nulling indicator lights (or a meter) connected to read the difference between the automatically and manually generated dc voltages for each function (Fig. 12-2). Once level match is achieved, the engineer can switch from read to write on a function without the levels changing. After the write mode is entered, the controls may be moved from the level match position to the new settings desired.

Update Mode

The *update mode* eliminates the need to match the manual settings to the automatic one when punching into an existing data track. In this mode, settings are changed by adding to or subtracting from the automatic control voltage rather than completely rewriting them. The advantage to this is that if some involved level changes were made correctly, except for the loudness and softness of the track, the overall volume could be changed without having to perform all the other level changes. The changes are reconstructed from the previous data track, raised or lowered in level by the manual update control, and recorded on another data track. A level match between the new and old control voltages is achieved by setting the manual control to a position indicated by the manufacturer of the automated console to cause no increase or decrease when the update mode is entered. Once in the update mode, the control is operated to change the programming as desired and is then returned to the update level match setting to resume the levels of the previous mix. At this point, either the read mode can be initiated on this function or the update mode can be retained with the manual control left at the neutral point.

Since at least two data tracks are required in order to store and update a mix, only 22 tracks are left for audio on a 24-track tape. If a separate mono and/or stereo single mix are desired in addition to the stereo LP mix, even fewer tracks can be used for recording the music. The solution to this is to either provide for a larger number of

**Fig. 12-2.
MasterMix VCA
fader** *(Courtesy
Audio Kinetics,
Inc.).*

tracks than is needed for the music (leaving some tracks blank), or to synchronize a second audio recorder with the original master by recording an SMPTE sync track onto both the master and data tapes. This allows instruments to be recorded on all but one track of the music tape regardless of how many separate mixes are to be stored, with the added advantage of having the time code printed on the master tape should additional sync be desired at a later date.

Scanning

The programmer senses all of the control voltages used in the console and encodes them into a form which can be stored and decoded at a later date. The number of control voltages needed in an automated console makes it prohibitively expensive to monitor each control input on a continuous basis. Instead, the programmer samples the dc level applied to each control device in a preset order. Each device is sampled for the same amount of time, usually a small fraction of a second. This sequential sampling is called *scanning* and it produces a single channel of output data which can be broken down by the decoder to reproduce the original dc levels. The time required to scan all the control inputs in the console once is called the *scan rate* or the *updating rate*. As used here, the term *updating* refers to the repeated sampling by the encoder of control voltages, one time during each cycle, to create new data for storage. Updating also refers to the repeated modification of the reconstructed control voltages in response to each new cycle of data fed to the decoder.

The updating rate must be fast enough to provide smooth and accurate encoding of the engineer's actions. For example, if the engineer fades out a level control at a rate of 5 dB per second and the encoder scans the corresponding control voltage only once each second, the programmer will sense and encode 5-dB steps rather than a continuous fade (Fig. 12-3A). If the scan rate is increased to 50 milliseconds, the programmer would process steps of 0.25 dB, which more closely approximates the continuous fade (Fig. 12-3B). The decoder output can be filtered to remove these small steps from the reconstructed dc voltage so that level changes are smooth. The higher the scan rate, the less filtering is needed because very small steps are perceived as a continuous change. The scan rate is limited by the storage capacity of the memory unit as well as by the time needed for the processor to encode or decode a voltage.

Fig. 12-3. A 5-dB-per-second fade of 1-second duration.

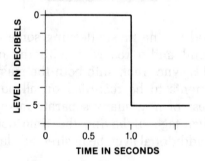 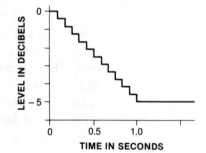

(A) Updating occurs once each second. The fade is encoded as a step change of 5 dB.

(B) Updating occurs every 50 milliseconds. The fade is encoded as 20 steps of 0.25 dB each.

The programmer processes two different types of functions. *Switching functions* have only two states and can be used to turn signals on or off (Fig. 12-4A), where continuous control is not required. A *dynamic function* controls devices that can be set to any position

between a maximum and a minimum, such as program fader level
(Fig. 12-4B).

**Fig. 12-4. Two
functions that a
programmer
processes.**

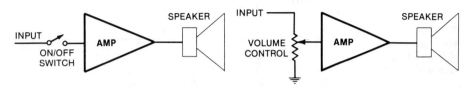

(A) A switching function.   (B) A dynamic function.

Encoding and Decoding

Since the signals being scanned are dc voltages, they must be
encoded into another format before they can be recorded onto a data
track. Two methods of encoding are currently in widespread use.
The first method encodes the analog dc levels as *digital binary words*.
Since digital words by their very nature can express only step
changes rather than continuous changes, switching functions are
encoded perfectly, while the accuracy of the encoding of dynamic
functions depends upon the size of the steps. Expressed in decibels,
the step size is called the *resolution* of the processing.

For now, convenience and cost require that the digitally encoded
words be recorded on an audio recorder or on a suitable digital memory device, such as a synchronized floppy disk. The *bit rate* (i.e., the
number of bits created by the encoder per second) cannot exceed the
storage density of the data track's recording capabilities. In other
words, the frequency bandwidth of the recorder must be equal to, or
greater than, the bit rate. For a rate of 9600 bits per second, a data
channel bandwidth of about 10 kHz is needed—a figure well within
the analog and digital storage medium capabilities. The code is usually set up so that step size becomes larger when control settings are
not as critical. For example, at attenuation settings below − 70 dB,
the steps may be as large as 2 dB each, while they may be just 0.25 dB
at a − 10 dB setting. Thus, bits are not wasted on settings which are
so low that they can barely be heard; they need to be present only to
provide a smooth fadeout. Since switching functions are either on or
off, they require only one bit of storage; thus, a greater quantity of
them can be automated over dynamic functions.

Each bit of data is represented by an electrical pulse, resulting in
a magnetic flux pulse on the data track. In order to ensure accurate
digital processing, these pulses must have rapid *rise* and *fall times*.
Fig. 12-5 illustrates that the rise time of a pulse is the time needed for

the level to increase from 10% to 90% of the maximum value and the fall time is the time needed to decrease from 90% to 10% of the maximum.

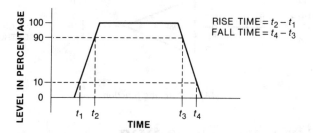

Although the encoder generates pulses of the proper shape, any nonlinearity of the electrical phase response in the tape recorder used for storage can distort this shape. To prevent possible errors, the decoder contains a *digital signal conditioner* which generates fast rise and fall time pulses from the distorted pulses on the tape before the data is decoded into analog form (Fig. 12-6). The generation of new pulses also prevents the errors that would result if tape distortion and noise were allowed to build up as a result of bouncing mix information back and forth between data tracks.

Using magnetic tape as the storage medium subjects the data to the problems of dropout and magnetic discontinuity at splices. Since the programmer depends upon the sequence of the recorded data to determine what corresponds to which function, a temporary loss of signal at the input of the decoder due to a dropout or a splice could cause an error in the assignment of control voltages. To prevent this, a *parity checker* is included in the programmer which senses *error signals*, such as the lack of input to the decoder, or noise which could otherwise be misinterpreted as data. The parity checker causes the programmer to hold all levels at values determined by the last complete scan. The length of time that the program can hold previous values while awaiting valid data is called the *error signal holding time* and is expressed as the number of decibel change in the control settings per second during a data dropout. When valid data is again received by the decoder, control voltage update resumes.

The automated mixdown system available in the Sony/MCI JH-600 recording console uses a biphase digital recording format which is entirely housed and integrated into the existing console mainframe. This automation system, as with most automation systems available today, has automation control over the individual channel fader level, the master fader level, the channel subgrouping, and the individual/group muting functions.

The individual fader controls (Fig. 12-7) of the JH-600 console affect only the input strip on which they are located, unless the channel is assigned as a group master. In this case, the controls will affect all channels assigned to that group. The *mute* button, located at the top of the fader, mutes the channel or group unless the solo-in-place function is programmed, in which case, the mute button is redefined as a *solo* button. This allows an easy function changeover between the need for *solo* in the recording process and the need for *muting* in the mixdown mode. The *mute write* button makes the channel or group mute an automated function. The *null indicators* indicate which direction to move the fader to achieve the null position. These are then followed by the *write* and *update* buttons, with *read* being the normal console status in the automation mode.

Grouping can be made with any number of channels; they are assignable by use of the thumbwheel switch on each channel, where group numbers 1 through 8 may be selected. Group masters are assigned by pressing the master switch on the channel selected, allowing these faders to control all relative levels assigned to this master group fader.

The JH-600 automation master fader (Fig. 12-8) supplies the overall dc voltage to the console's individual faders and, as such, is the automation console master. Located on the panel of the fader are three master function switches. The *master mute write* button switches all channels into the mute write mode, while the *fader write* and *fader update* selectors place all the console channels into the write or update modes respectively.

The JH-600 automation master control panel (Fig. 12-9) houses four selectable master functions. By pressing the master *clear (CLR)* button, all functions programmed by the automation control buttons are placed into the read mode. Pressing the button a second time within 5 seconds sets all channel VCAs to approximately -12 dB

Fig. 12-7. The JH-600 automated VCA fader control panel.

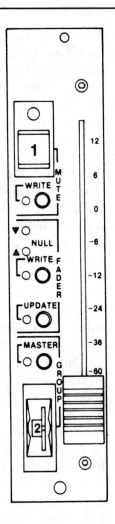

and cancels all channel mutes. This clears the console for a remix session. The *memory map (MEM)* button is used to write automation data into a random-access memory (RAM) in a scratchpad fashion, rather than onto a data track, for experimentation with a mix. The data may then be recorded onto the working data track (in the update mode) upon attaining the desired results. The *LOCK* button places the console in the read mode and acts as a safety lockout switch, preventing the accidental pressing of any automation function switch when recording the initial mix onto the data track. The *SOLO* selector switches the functions of all the individual channel fader mute buttons into that of a solo button.

Since the majority of automation data signals are encoded in digital form, Audio Kinetics Inc. has introduced their MasterMix MX644

Fig. 12-8. The JH-600 automation master fader panel.

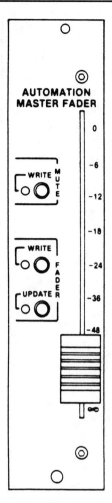

System (Fig. 12-10), an independent automation system for audio consoles. This system stores all automation information digitally onto a double-sided quad-density 5.25-inch floppy disk recording system. Synchronization is maintained by way of SMPTE time code on both the MasterMix and the multitrack master recorder, eliminating the need for multitrack data tracks and assuring SMPTE on the master for future synchronization. Equipped with an SMPTE time code generator, it offers 25-frame, 30-frame, and drop-frame formats. Comprised of a central processing unit, a floppy disk drive, and a keypad control unit, the MasterMix system will store and update up to four separate automated mixes. It is capable of 600K bytes of storage using a standard 5.25-inch computer diskette. Archive safety copies of important mixes may be easily made onto a separate floppy disk.

**Fig. 12-9. The
JH-600
automation
master control
panel.**

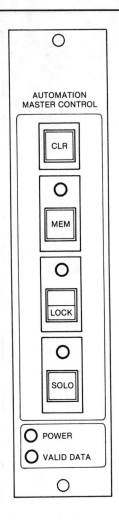

**Fig. 12-10.
Audio Kinetics'
MasterMix
MX644
automation
system** *(Courtesy
Audio Kinetics
Inc.).*

This system is designed to be easily incorporated with a console that offers dc or digitally grouping faders. Older consoles that are not equipped with automation may be fitted with a console interface when used in conjunction with MasterMix faders.

References

1. Bloom, Robert A., Letter reprinted in *Recording engineer/ producer*, Vol. 3, No. 4, July/August 1972, p. 24.
2. Buff, Paul C., Robert A. Bloom, Saul Walker, Daniel N. Flickinger, Claude Hill, Wayne Jones, and Ron Neilson; Individual letters reprinted as "Automation for Audio Recording," *Recording engineer/producer*, Vol. 3, No. 4, July/ August 1972, pp. 23-32.
3. Massenberg, George, "Parametric Equalization," *Audio Engineering Society*, Preprint No. K-2, 42nd Convention, May 2-5, 1972.
4. *MasterMix MX-644 Technical Bulletin*, Audio Kinetics Inc.
5. *NECAM 96 Technical Bulletin*, Rupert Neve, Inc.
6. "Sony/MCI Automation System," *JH-600 Operating Manual*, Sony/MCI Corp.
7. Talks by Saul Walker and Paul Buff to the New York section of the Audio Engineering Society on November 18, 1972.
8. Yentis, Wayne, "A Glossary of Automation Terms," *Recording engineer/producer*, Vol. 3, No. 6, November/December 1972, pp. 29-31.

13 SPEAKERS AND MONITORS

In the recording process, both judgments and adjustments of sound quality are based entirely on what is heard through the monitor system; thus, it is extremely important that monitors be set up and operated properly. Speakers are the weakest link in the audio chain due to the fact that their response curves are difficult to make flat. In addition, the acoustics of a room can create large peaks and valleys in the frequency response at the listening position. The only place a speaker can truly be designed to have a flat response is in an anechoic chamber, i.e., a room that absorbs all the speaker output and reflects none of it back. In such a room, there can be no constructive or destructive interference of the reflected waves and what is heard is the direct output of the speaker. Since people do not listen to music in anechoic chambers, the listening room must be taken into consideration when choosing a speaker.

Speaker/Room Considerations

Unless the rooms are of identical dimensions and furnishings, a speaker will sound different (it will have a different frequency response curve) in every room in which it is placed. This means the speakers must be tested in the room in which they are intended to be used. Direct comparisons of various speakers in the same room are valid only in demonstrating differences between the speakers; direct comparisons cannot determine which one will sound best in another room. This problem of sound variation from room to room makes it difficult to interchange recording studio control rooms. Even if high standards of acoustic construction and tuning are followed, no two rooms will sound exactly alike. After a tape is recorded in a specific control room, the producer and artists become accustomed to hearing the material sound a certain way. If the tape is mixed down in another control room with the same speakers but with a different speaker/room response curve, there can be quite a difference in the sound of the instruments.

To eliminate this, many studios *tune* their speakers to the room so that the frequency response curve of the speaker/room is reasonably flat and, therefore, compatible with most other control rooms. The tuning is accomplished using the peaking and dipping filters of a graphic equalizer (Fig. 13-1) which is usually of a ⅓-octave bandwidth and connected before the monitor power amp. Pink noise (which has a flat energy spectrum curve throughout the audio range) is fed into the speaker system. Then, the acoustic outputs are measured one at a time in ⅓-octave increments using an instrument known as a spectrum analyzer (Fig. 13-2). This is a device which will give a visual display of the frequency response as measured through a specially calibrated omnidirectional microphone. The spectrum analyzer will then give an instantaneous reading of the frequency response of the room at the microphone's specific location.

Fig. 13-1. The UREI Model 535 dual graphic equalizer.

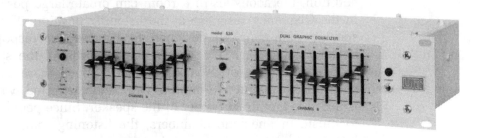

Fig. 13-2. Klark Teknik DN 60 Real-Time Spectrum Analyzer (*Courtesy Klark Teknik Research Ltd.*).

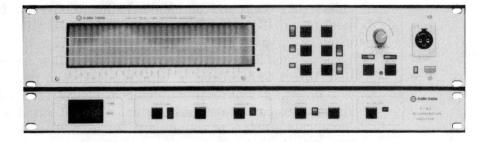

An omnidirectional mike is used because the ear is omnidirectional and thus hears the reflected sound of the room as well as the direct sound from the speaker. Since the mike can only be at one spot at any one time, the response curve obtained will be accurate only for a listener at the spot where the mike was placed. Although modern control room design is improving sound distribution, the response curve of a tuned system will vary from one spot in a room to another. The response at the engineer's and producer's seats must often be a compromise so that both hear similar responses.

Examples of equalizers used in tuning speaker systems are the UREI Model 537, the Klark Teknik DN-360, and the White 4200-A graphic equalizers. Pink noise is used for testing rather than sine waves, because it is of a random nature and does not stimulate standing waves in a room as a sustained tone would. The presence of standing waves would introduce inaccuracies in the readings on the analyzer that would vary with the position of the microphone in the room.

Although equalizers can flatten speaker response, the output of many speaker systems and amplifiers falls off at the low end and becomes distorted if too much low-frequency signal is applied. Thus, flattening the response of a speaker system may lower its undistorted volume-producing capability, making it unusable if high monitor levels are desired.

Speaker Design

Just as one speaker will sound different from another in differing acoustic environments, so will speakers of differing designs vary widely in sound character. Enclosure size, number of drivers and their size in each enclosure, crossover frequencies, and design philosophy will contribute to differences in sound quality. Speaker enclosures may be of two design types: air suspension and bass reflex.

An *air-suspension* speaker enclosure is an air-tight sealed system, separating the air in the enclosure's interior from the air outside the enclosure. This system improves the bass response of small speaker systems whose enclosures would otherwise not be large enough to effectively reproduce bass frequencies (Fig. 13-3A). The *bass-reflex* system requires a larger size speaker enclosure in order to reproduce the lower frequencies. In this design, a *bass* port hole is cut into the face of the speaker enclosure, allowing the air mass inside the enclosure to mix freely with the air outside the enclosure (Fig. 13-3B).

With so many variables to consider in speaker/room arrangements, there is no such thing as the "ideal" monitor speaker system. It is more a matter of taste. Those monitors that are widely favored over a long period of time gradually tend to become regarded as the "industry standard," but this can easily change as preferences vary.

**Fig. 13-3.
Speaker
enclosure
designs.**

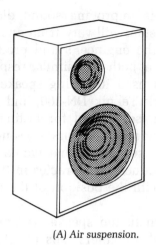

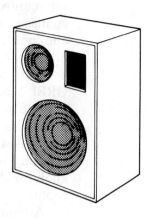

(A) Air suspension.   (B) Bass reflex.

Crossover Networks

Because individual speaker elements (called drivers) are more efficient in some frequency ranges than in others (i.e., have more undistorted output for the same input level signal), different drivers are often used in conjunction with one another to obtain the desired output. Large-diameter drivers, such as 15-inch units, produce low-frequency information more efficiently than high-frequency information, while medium-size speakers, such as the 4- and 5-inch units, produce midrange frequencies better than the highs or lows, and small speakers (½- to 1-inch size) produce highs better than any other range.

These speakers are connected by *crossover networks*, which prevent any signals outside a certain frequency range from being applied to the speaker. The networks usually have one input and two or three outputs. Input signals above the crossover frequency are fed to one output, while signals below the crossover frequency are fed to another. The crossover network uses inductors and capacitors and is designed so that a signal at the crossover frequency will be sent equally to the respective outputs, providing a smooth transition from speaker to speaker. If a speaker system has only one crossover frequency, it is called a *two-way system* because it divides the signal into two bands. If the signal has two crossover frequencies, it is called a *three-way system*.

The JBL 4430 monitor speaker used in many recording studios is a three-way system. It utilizes a 15-inch woofer for the bass and a horn-type driver for the midrange and high frequencies. The lower crossover frequency is 1000 Hz, and a level control determines how much energy is sent to the middle- and high-frequency drivers from the crossover network. This control enables the user to partially compensate for differences in room acoustics. An absorptive room will require more high-frequency energy than a live room to produce the same audible effect.

Better crossover networks, called *electronic crossovers*, differ from conventional crossover systems in that they are used between a preamp and several amplifiers instead of being connected between a single power amp and several drivers. Each speaker driver is fed directly by its own power amp (a three-way system would need three power amps per channel). There are several advantages to this approach:

1. Since the signals are at low levels within the electronic crossover, active filters without inductors can be used, thus removing a source of intermodulation distortion.
2. Power losses due to the resistance of inductors in the passive crossover network are eliminated.
3. Since each frequency range has its own power amp, the full power of the amplifier is available to it regardless of the power requirements of the other ranges.

For example, given a 100-watt amplifier feeding high- and low-frequency range drivers through a passive crossover network: If the low frequencies are using 100 watts of power, and a high-frequency signal comes along that requires an additional 25 watts of power from the amp, the amplifier cannot supply it, and both the low- and high-frequency signals become distorted.

To illustrate further, picture a two-way system with 8-ohm drivers and an 8-ohm passive crossover network (Fig. 13-4A). The program material requires that the power amp be capable of supplying 100 watts of power to the low-frequency speaker and 25 watts of power to the high-frequency speaker at the same time. Producing 100 watts in 8 ohms requires 28.3 volts at the speaker terminals, while 25 watts requires 14.1 volts (voltage $= \sqrt{\text{power} \times \text{resistance}}$). This means that the power amplifier must feed 42.4 volts to the crossover network (assuming a no-loss network), which is a power output of 225 watts. The system requirements could also be met through the use of a 100-watt amplifier to drive the low-frequency speaker and a separate 25-watt amplifier to drive the high-frequency speaker, with the input signal fed to the power amps through an electronic crossover network (Fig. 13-4B). Systems using electronic crossovers and multiple-power amplifiers are called *bi-* or *tri-amplified systems*, depending upon the number of power amps used per channel.

Fig. 13-4.
Passive and
bi-amplified
crossover
systems.

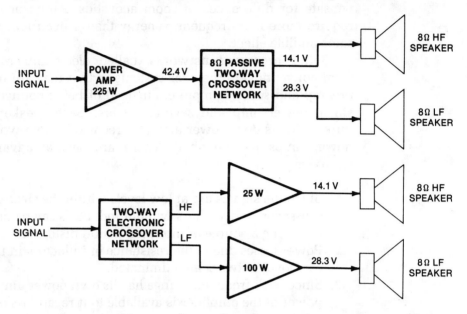

Passive and active crossover networks are similar in their frequency responses. The crossover point is always 3 dB down from the flat section of the response curve; slopes outside the filter passband are usually 6, 12, 18, or 24 dB per octave with 12 being the most common (Fig. 13-5). Common crossover frequencies are 500 Hz, 800 Hz, 1200 Hz, 5000 Hz, and 7000 Hz.

Speaker Phasing

A pair of speakers can be in phase or out of phase with each other. If they are in phase (Fig. 13-6A), the same signal applied to both speakers will cause their cones to move in the same direction. If they are out of phase (Fig. 13-6B), one cone will move inwards while the other will move out. Phase can be tested by applying a mono signal to the two speakers at the same level. If the signal appears to originate from between the speakers, they are in phase, but if the image is hard to locate and appears to move as the listener moves his head, they are out of phase. This effect is especially noticeable with low frequencies. An out-of-phase speaker condition is corrected by reversing the leads connected to one of the speakers.

Speakers are phased according to a certain code: a 1½-volt battery is connected to the speaker so that the cone moves outward. The

**Fig. 13-5.
Frequency
response of a
3-way crossover
network with
crossover
frequencies of
500 Hz and 5
kHz.**

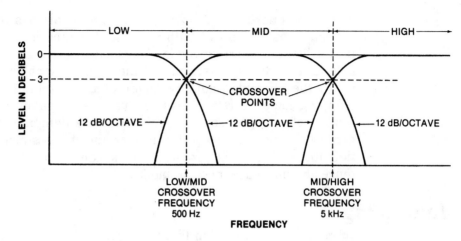

**Fig. 13-6.
Relative cone
motion;
speakers in
phase and out of
phase.**

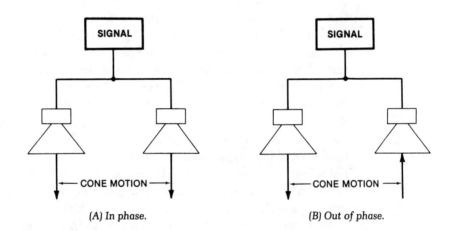

(A) In phase.   *(B) Out of phase.*

speaker lead to which the positive side of the battery is attached is then marked with a " + " or a red dot. *Caution:* This test should not be attempted on high-frequency tweeters because it may damage their diaphragms or voice coils.

Lamp cord is often used for speaker connections, although heavier-grade or multiple-strand low-resistance speaker cable is available. On either of these cables, there will often be an identifier mark on one side of the line. If this coded-wire side is always connected to the " + " or red dot of the speaker and to the high side of the power-amplifier output, speaker-phase problems will be avoided.

Speaker wire should always be heavy duty; No. 18 wire is the proper size for less than 25-foot lengths, No. 16 for 25- to 50-foot lengths, and No. 14 for 50- to 100-foot lengths. The reasons for increasing the thickness of the conductor as cable length increases

(No. 16 is thicker than No. 18) are: (1) all cable has resistance, and resistance builds up as length increases. The more resistance there is in a cable, the more power is dissipated in the cable and is unavailable to drive the speaker; and (2) the higher the cable resistance, the lower the effective damping factor of the amplifier. The amplifier damping factor is related to how well the amplifier can control the motion of the speaker cone. The lower the damping factor, the less control the amp has over the speaker and the less clarity and definition the speaker will produce. Thick conductors have lower resistance and minimize these problems.

Monitoring

In mixing, it is important that the engineer be seated exactly between the stereo speakers and that their volume be adjusted to be equal. If this is not done, signals desired in the center of the speaker sound field may be off to one side or the other. If the engineer is closer to one speaker than the other, that speaker will seem to be too loud and the other speaker too soft. The engineer may be tempted to either pan the instruments towards the far speaker or boost that entire side of the mix to equalize the volumes. The mix will sound centered in that specific control room but when the mix is listened to in another environment, the mix will be off center. As a check against this, the engineer should always make sure that an audible volume difference between speakers is accompanied by a corresponding visual difference on the VU meters monitoring the signal sent to tape. Another guard against off-center levels is to check the levels from each individual monitor by placing a microphone in the center listening position (Fig. 13-7) and reading the outputs on meters or a VU meter at a spare input, allowing the signal outputs to be accurately matched up. While the meters should not read exactly the same at all times, the presence of a solo in one channel should read only a few dB higher than the other channel unless the other channel is being kept very low for a specific purpose. The maximum readings on the meters should be about the same. Center channel balance may be best checked by switching on the console's line-up oscillator, which allows the main output pair to be calibrated on the stereo output meters for identical left and right output levels.

Mixing

Several other problems remain to be considered with respect to monitoring. Even if the monitor speakers in use are absolutely flat in the control room, few of the people who buy records have flat speaker/

Fig. 13-7. Balancing the center monitoring image in a studio may be accomplished by placing program material or pink noise into each monitor speaker and measuring the output of each with a centrally placed microphone which is fed to a channel monitored by a VU meter. The relative outputs of the speakers are adjusted until they are equal.

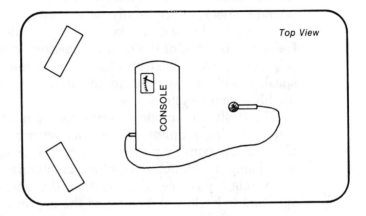

room curves and, as a result, will not hear exactly the same mix that is heard in the control room. Purchasers will hear different frequency balances due to response variances between types of speakers and listening rooms.

There are several possible views on this matter:

1. Mix for those few people who have flat systems; make the rest of the purchasers accept whatever inaccuracies are introduced by their systems.
2. Mix on home-type hi-fi speakers of medium quality; compare the sound of the mix on several different brands of speakers, adjusting the mix until it sounds good on as many systems as possible.
3. Mix through inexpensive speakers, such as those found in car radios and some portable record players.

The problem with the first view is that owners of flat systems are not the majority of record buyers. Catering to this market, while neglecting the larger group of people who buy most of the records, will result in poor sales. In the second view, mixing for the consumer with medium-quality speakers will please the majority of record buyers, but not the artist or producer. The third view of mixing through inexpensive speakers applies to special markets, such as a-m radio or teen-age music, where the song will receive most of its play through low-quality equipment and its balance must be set so that the music sounds good on this equipment. Car radio speakers and cheap phonograph speakers are notorious for their lack of bass and highs but with an abundance of peaky midrange response. Deficiencies in the frequency balance of the monitor speakers are inverted when the mix is played back on flat speakers; i.e., if the mix sounds good on a speaker with a big peak in the midrange response, it will sound muddy on a flat system.

Generally, in a modern recording studio, a combination of two or three differing monitor systems, as described above, will be selectable from the console position. Each will generally have its own associated amplifier which allows for a greater degree of level matching.

Another possible view is that the sound of the control-room speakers is irrelevant as long as the engineer knows how the speakers must sound in order for the mix to sound good on other speakers. If there is a midrange peak in the control-room speakers, the engineer can purposely make the mix sound shrill and piercing because he knows that on flatter speakers the excess midrange will not be present and the mix will not sound shrill. While this approach works in practice, it is very difficult to convince producers and artists that the mix must sound bad at the studio in order to sound good at home.

A compromise solution is to tune the monitor speakers flat up to 5 or 8 kHz and then gradually roll off the response so that it is down about 10 dB at 16 kHz. This will make the engineer boost the level of the high frequencies in the mix. Since almost all consumer hi-fi equipment is deficient in high-frequency response, the boost is rolled off in the listener's home. The boosted high end has become a standard in the pop music industry over the past decades so that the high-frequency rolloff in the monitor speakers will make records sound the way that people are used to hearing them.

Some typical speakers and studio monitors are shown in Figs. 13-8 through 13-12.

Monitor Volume

To make monitoring even more difficult, the Fletcher-Munson curves come into play, making the frequency balance of a mix vary depending upon the monitor volume. If the balance is set while listening at loud levels and then the mix is played back softly, the bass and extreme highs disappear. If the balance is set while listening to softly played sounds and the mix is played back loudly, the bass and

**Fig. 13-8. The
Auratone Model
5C Super Sound
Cube™. Speaker
on the right has
its black foam
grille removed**
*(Courtesy
Auratone Corp.).*

**Fig. 13-9. The
JBL Model 4312
studio monitor**
*(Courtesy JBL
Inc.).*

**Fig. 13-10.
Model 4430 and
Model 4435
studio monitors**
*(Courtesy JBL
Inc.).*

Fig. 13-11. The UREI Time Align Crossover Network *(Courtesy JBL Inc.).*

extreme highs become excessive. Since it is impossible to know how loudly a listener will play the song, it is difficult to know how loudly to monitor while mixing. It appears that 85-dB spl is the best monitoring level from the standpoint of minimum change in apparent frequency balance that is due to a change in playback loudness and softness. Conveniently, average home listening levels are in the 75- to 85-dB spl range. The response that is considered flat at 120-dB spl changes when it is played back at lower levels. As playback volume decreases, the bass first increases and then begins to decrease, and output in the presence range begins to fall. So, a mix made at 120-dB spl will sound distant and lifeless at lower levels. If the mix is made at 100-dB spl, dips occur in the presence range as monitor volume is decreased but the most noticeable change is at the low end, where 50 Hz is 12 dB down at 70-dB spl. If mixed at 85-dB spl, however, playback between 90- and 60-dB spl causes response changes which are less than 5 dB at the extremes of the spectrum and practically no change within the spectrum. Thus, this is a good compromise level for mixing, allowing for a certain amount of variation in playback level by consumers without adversely affecting the sound.

Fig. 13-12. Some Auratone quality sound monitors. Clockwise from bottom left is the Model T66, Model T5, Model RC66, Model T6, and Model QC66 *(Courtesy Auratone Corp.).*

Compatibility

Another concern in monitoring is mono/stereo compatibility. Some fm and all a-m radio listeners listen in mono, and radio introduces much of the new pop music to the consumer. Thus, if an album sounds good in stereo but poor in mono, it may not sell well because its greatest initial exposure will be in mono. To prevent this, mixes should also be listened to in mono to make sure that no out-of-phase components are present which would cancel out instruments and degrade the balance.

References

1. Crowhurst, Norman H., "Theory and Practice," *db, The Sound Engineering Magazine*, Vol. 3, No. 6, June 1969.
2. Eargle, John, "Equalizing the Monitoring Environment," *Journal of the Audio Engineering Society*, Vol. 21, No. 2, March 1973, pp. 103-107.

3. Eargle, John, and Mark Engebretson, "A Survey of Recording Studio Monitoring Problems," *Recording engineer/producer*, Vol. 4, No. 3, May/June 1973, pp. 19-29.
4. Malo, Ron, "Monitoring . . . You Can't Hardly Believe What You Hear!" *Recording engineer/producer*, Vol. 2, No. 3, May/June 1971, pp. 23-24.
5. Siniscal, Albert, "Bi and Tri Amplification," *Recording engineer/producer*, Vol. 2, No. 2, March/April 1971, pp. 27-29.
6. Small, Eric, "Problems of Mono-Stereo Broadcasting Compatibility," *db, The Sound Engineering Magazine*, Vol. 5, No. 8, August 1971, pp. 26-29.
7. Smith, David, D. B. Keele, Jr., and John Eargle, "Improvements in Monitor Loudspeaker Systems," *Journal of the Audio Engineering Society*, presented at the 69th Convention, May 12-15, 1985.

14 PRODUCT MANUFACTURE

Once a tape has been approved in a mixdown master-tape form, the next step is to transform this master tape into a saleable mass-producible form. With the present available technology, that product may show up in the form of either a record, a cassette tape, or a compact disc (CD). Each of these products has a specific manufacturing process and method of quality control; careful attention is required throughout the entire manufacturing process.

Disc Cutting and Pressing

The first stage of record production is the disc cutting process. As the master tape is played on a specially designed tape playback machine, its signal output is fed through a *disc mastering console* to a *disc cutting lathe*. Here the electrical signals are converted into the mechanical motion of a stylus and are cut into the surface of a lacquer-coated recording disc.

Unlike tape which maintains the chronological order and duration of recorded information by relating periods of time to physical lengths of tape, discs relate periods of time to the angle of disc rotation (Fig. 14-1). As the turntable rotates at a constant *angular velocity*, such as 33⅓ or 45 rpm, the stylus gradually moves closer to the disc center, cutting a continuous spiral into the disc surface. The time relationship of the recorded material can be reconstructed by playing the disc on any playback turntable that has the same constant angular velocity as the one used to record the disc.

The system of recording used for stereo discs is the 45/45 system. The recording stylus cuts a 90° angle groove into the disc surface so that each wall of the groove forms a 45° angle with the vertical. Left-channel signals are cut into the inner wall of the groove and right-channel signals are cut into the outer wall (Fig. 14-2A). Fig. 14-2B is a photomicrograph of grooves showing recorded stereophonic mate-

**Fig. 14-1. A
period of time is
represented by a
disc rotation of
angle θ.**

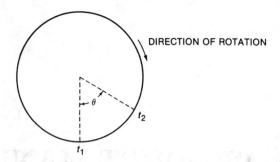

Fig. 14-1. A period of time is represented by a disc rotation of angle θ.

rial. The stylus motion is phased so that a signal which is in phase in both channels (a mono signal or a signal centered between the two channels) produces lateral motion of the groove (Fig. 14-3A); out-of-phase signals (channel difference information) produce vertical motion (Fig. 14-3B), i.e., changes in groove depth. Because this system is compatible with mono disc systems, which use only lateral groove modulation, a mono disc can be accurately reproduced with a stereo playback cartridge. However, unless monophonic cartridges are designed for stereo compatibility, they will reproduce only lateral motion and resist movement in the vertical direction. This resistance will damage the stereo information contained in the vertical motion of the groove if a stereo record is played with a mono cartridge.

Cutting Lathe

The main components of the modern disc cutting lathe are the turntable, the lathe bed and sled, the pitch/depth control computer, and the cutting head. The Neumann VMS-80 lathe is illustrated in Figs. 14-4 through 14-7. The tube entering the front of the cutter head provides helium cooling to permit extended operation at high cutting velocities.

The turntable is very heavy in order to reduce speed variations via the flywheel effect. It is driven by a special motor and linkage system to eliminate flutter and rumble from the recording. Three sets of stroboscopic rings on the outer rim calibrate the four switch-selectable speeds of the turntable: $16\frac{2}{3}$, $22\frac{1}{2}$, $33\frac{1}{3}$, and 45 rpm. A vacuum suction system secures the recording blanks to the turntable via holes in the turntable surface. The holes can be selectively opened or closed to provide proper suction to hold lacquer discs from 7 to 16 inches in diameter. The suction is introduced through a flexible pipe connected to the hollow center post of the turntable.

Cutting Head

The *cutting head* (Fig. 14-7) translates the electrical signals applied to it into the mechanical motion of the recording *stylus*. The stylus

**Fig. 14-2.
Illustrations of
stereo record
cutting.**

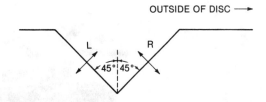

OUTSIDE OF DISC ⟶

(A) Left-channel channel signals are cut perpendicular to the inside groove wall, while right-channel signals are cut perpendicular to the outside groove wall.

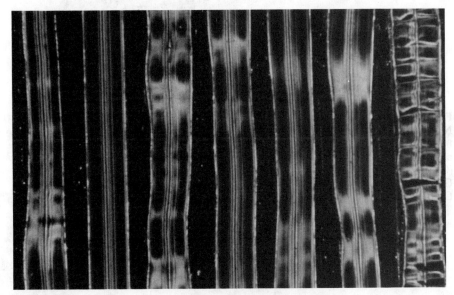

(B) A photomicrograph of grooves modulated with stereo program material.

**Fig. 14-3.
Groove motion
in stereo
recording. The
solid line is the
groove with no
modulation.**

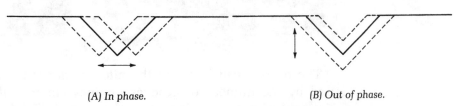

(A) In phase.   *(B) Out of phase.*

gradually moves in a straight line toward the center hole of the disc as the turntable rotates, creating a spiral groove on the surface of the record. This motion is achieved by attaching the cutting head to a *sled*. A spiral gear known as the *lead screw* drives the sled in a straight track called the *lathe bed*. The speed of the cutting head motion towards the center of the disc determines the playing time of that side of the record.

**Fig. 14-4.
Neumann VMS-
80 disc cutting
lathe** *(Courtesy
Gotham Audio
Corp.).*

**Fig. 14-5.
Neumann VMS-
80 lathe pitch
controller**
*(Courtesy
Gotham Audio
Corp.).*

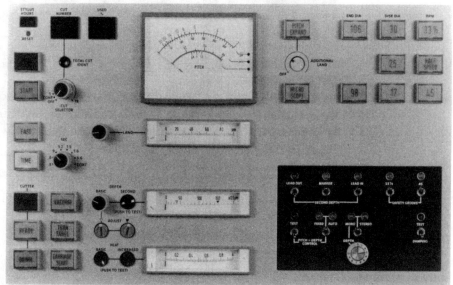

The head speed is called the *pitch* of the recording and is measured by the number of grooves or *lines per inch (lpi)* cut into the disc. As the head speed increases, the number of lpi decreases, so the pitch and playing time also decrease. Several methods of changing pitch are possible: the lead screw can be changed for one with a finer or coarser spiral, the gears that turn the lead screw can be changed to change the speed of the lead screw rotation, or the lead screw rotational speed can be varied directly by varying the speed of the motor driving it. The latter method is used in the Neumann lathe and provides a continuously variable pitch.

**Fig. 14-6.
Neumann lathe
bed and sled**
*(Courtesy
Gotham Audio
Corp.).*

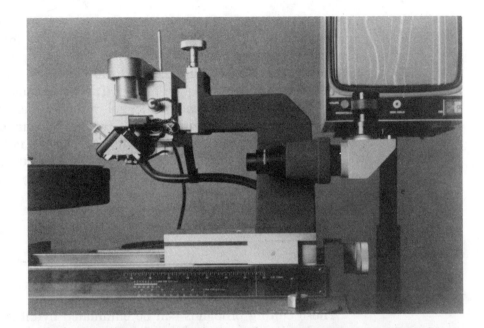

**Fig. 14-7.
Neumann SX-68
cutter head**
*(Courtesy The
Master Cutting
Room).*

The space between grooves is called a *land* (Fig. 14-8). Unmodulated grooves are equally spaced at all points. Adding modulation to the grooves produces a lateral motion proportionate to the in-phase signals contained in the two channels being cut. If the cutting pitch is too high (too many lines per inch, making the grooves very closely spaced) and high-level signals are cut, it is possible for the groove to break through, or *cutover*, the wall of an adjacent groove, or for the grooves to overlap, which is called *twinning*. The former is likely to cause the record to skip when played. The latter causes either distortion of the signal or an echo of a signal in the adjacent groove, due to the deformation of one groove wall by the information cut in the next. Groove echo can occur even if the walls do not touch. It is a function of groove width, pitch, and level, and it decreases as the signal frequency increases. In addition, high-frequency echoes decrease in level as groove diameter decreases.[4]

Fig. 14-8. The land portion of a recording.

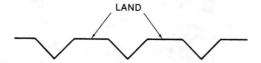

These cutting problems can be eliminated by either reducing the cutting level or by cutting fewer lines per inch. A conflict arises here because, in comparison with a softer record, a louder record sounds brighter, punchier, more present, and fuller. As a result, record companies and producers are concerned about the *competitive level* of their discs relative to those cut by others, so they do not want to reduce the cutting level. However, reducing the pitch shortens the playing time of the record.

The solution to these problems is to vary the pitch, cutting more lines per inch during soft passages and fewer lines per inch during loud passages. This automatic pitch control is achieved by adding an additional *preview playback head* to the tape machine feeding the lathe. The machine in Fig. 14-9 has such a head, which is positioned at a distance ahead of the regular playback head on the tape machine. This gives the *pitch/depth control computer* in the lathe, which determines the pitch required for each portion of the program and varies the speed of the lead screw motor, time to change the pitch as required. Since left-channel signals are cut in the inside groove wall and, therefore, run no danger of cutting into an existing groove, a pitch change is not needed for loud left-channel signals until the moment they are cut. At that point, the pitch is decreased and the grooves expanded so that the following groove will be far enough away that a cutover cannot occur. Since the pitch correction is not needed until the signal is cut, the computer derives its level information from the left-program channel. Right-channel signals, however, are cut in the outside wall of the groove, so loud right-channel signals require that the pitch change occur before the signal is cut, to make room for the new groove so that it does not cutover into the preceding groove. In order to provide the computer with right-channel signal-level information before the signal is cut, a *digital delay line* or a *preview playback head* with its associated playback electronics must be added to the tape machine. A very-high quality digital delay is sometimes used, such that the undelayed signal feeds the computer with information while the delayed signal is actually the signal fed to the cutting head. For the most part, an analog preview head is used. This head is positioned at a distance ahead of the program playback head of the machine (16.5 inches for a 33⅓-rpm disc and 15-ips tape). While pitch control requires preview information only for the right channel, *depth control*, to be described later in this chapter, requires preview information from both the left and right channels, so a stereo preview head is used. When cutting a mono disc from a stereo tape, the sum of the left and right preview channels is used for pitch control information. The computer samples the left program and the right preview signal-level information every one-quarter revolution of the turntable and adjusts the pitch to the value required by the highest of the current and the two previous level samples.

Fig. 14-9. Studer A80 master recorder which allows a 2½-inch preview for disc mastering.

Pitch is divided into two categories; *course*, which refers to between 96 and 150 lpi, and *microgroove*, which is between 200 and 300 (or more) lpi. Microgroove records have less surface noise, wider frequency range, less distortion, and greater dynamic range than course-pitch recordings. They can also be tracked with lower stylus pressure, resulting in longer life. This lower tracking force, however, makes the stylus more likely to skate across the record if the turntable is not level. The playback stylus for a stereo microgroove record must have a tip radius of 0.7 mil or less, as compared to 2.5 mils ± 0.1 for coarse-groove records. Early 33⅓-rpm and 78-rpm records were recorded with a coarse pitch. Virtually all current records are microgroove with 265 lpi being an average pitch. At maximum pitch, the playing time of one side of a 12-inch disc, with no modulation in the grooves, is 45 minutes. The duration of modulated 12-inch discs cut at average levels is 23 to 26 minutes per side when they are cut with a variable-pitch lathe.

Depth Control

In mono cutting, the depth of the groove remains constant since there is no difference between what would be the left and right channels of a stereo disc and, thus, there is no vertical information. The depth of the groove in a stereo disc varies with the vertical excursions of the cutting stylus and is measured from the surface of the disc to the bottom of the groove. If the depth is too great, the stylus will cut through the lacquer surface into the metal base of the recording disc, causing distortion and possible damage to the recording stylus. If the depth is too shallow, the cutting stylus could rise off the disc during a highly modulated passage and the groove would stop. If this disc were played, the playback stylus would skip at the point where the groove stopped and would jump either to another groove or completely off the disc.

Ideally, groove depth should not go below 2 mils for reliable tracking on all turntables. One mil is a standard compromise for the minimum depth needed to cut louder records. Grooves that are less than ¾ mil deep are considered too shallow or *light* to provide reliable tracking and are likely to cause skipping.

The problem of light grooves can be eliminated by either decreasing the separation between the channels when using a constant-depth lathe or by using a lathe with automatic depth control. This automatic control is achieved through the same pitch-/depth-control computer and preview playback head that was described earlier.

For depth control, the preview head outputs are added together out of phase to produce a signal equal to the upcoming information that is to be cut vertically. This signal is then applied to the depth-control amplifier. Since the stylus point forms an approximate 90° angle, deepening the groove will also make it considerably wider, so the pitch- and depth-control amplifiers are interconnected to expand the distance between the grooves when the cut is deepened. Thus, when strong out-of-phase or random-phase signals are present, the depth-control amplifier receives a greater signal than usual and deepens the cut to prevent the groove from becoming too light.

Recording Discs

The recording discs used on the lathe are very flat aluminum discs that are coated with a film of lacquer, dried under controlled temperatures, coated with a second film, and dried again. The quality of these discs, called *lacquers*, is determined by the flatness and smoothness of the aluminum base; any irregularities in its surface, such as holes or bumps, will cause similar defects in the lacquer coating. The disc flatness is achieved by stretching the aluminum. This can produce a *cosmetic effect* of two flashes of reflected light per revolution of the disc because the lacquer is not completely opaque. The presence of this effect does not degrade the quality or recording capability of the lacquer.

The lacquers used for *mastering* (cutting the lacquer to be sent to the pressing plant) are always larger in diameter than the final record, making it easy to handle the master without damaging the grooves. A 12-inch album is cut on a 14-inch lacquer, while a 7-inch single is cut on a 10- or 12-inch lacquer. Producers often cut a *reference lacquer* to hear how the master tape will sound after being transferred to disc. Long-playing references are cut on normal-size discs because most turntables cannot handle a disc that is more than 12 inches in diameter. The lacquers used for references are noisier and of poorer quality than those used for masters.

The recorded disc consists of several distinct sections as shown in Fig. 14-10: (A) the starting spiral, (B) the lead-in grooves, (C) the program, (D) the lead-out groove, (E) the spiral out, and (F) the locked groove.

The *starting spiral* is cut at a very low pitch of 6 to 10 lpi; it serves to catch the playback stylus as the stylus is lowered onto the record and feeds the stylus to the lead-in groove. Between one and three spiral-in grooves are recommended. This standard is especially important for a record changer since its tone arm falls at a preset distance from the center pin and the stylus must land in a spiral-in groove. If an insufficient number of spiral-in grooves are cut, the tone arm could fall outside the groove, either failing to feed into it or jumping off the disc.

The *lead-in groove* is unmodulated and is cut at the pitch preset as the maximum for the program material. This preset pitch, together

Fig. 14-10. The
different
grooves cut on a
disc.

with the starting spiral which feeds the stylus into the leading groove, stabilizes the tone-arm motion. The lead-in groove must be at least one complete revolution long. The first modulated groove is cut at a diameter no greater than 11$\frac{7}{16}$ inches for a 12-inch disc and 6$\frac{9}{16}$ inches for a 7-inch disc. The last modulated groove is cut at a diameter no less than 4$\frac{3}{4}$ inches for a 33$\frac{1}{3}$-rpm record and no less than 4$\frac{1}{4}$ inches for a 45-rpm record. The limitation on outer-groove diameter helps standardize the location of the starting spirals. The inner-groove limitation is a combination of standardization for the lift-off function for record changers as well as a prevention against severe high-frequency losses and distortion which result from the low groove velocities at small diameters.

Inner spirals or *bands* are often cut between sections of the program to facilitate finding different selections by reading the selection number on the disc label and counting the sections of program between the spirals. The spiral grooves used for banding can be either modulated or unmodulated and are cut at the same pitch as the starting spiral.

After the program grooves are cut, at least one unmodulated *lead-out groove* is cut before the *spiral-out* action begins. The final spiral is used to start the automatic lift-off and change cycle of record changers. This spiral leads into the last groove on the disc, called the *locked groove*. The locked groove leads back into itself and holds the stylus at the same groove diameter until the record-changer cycle begins. If the groove were not locked, the stylus might continue towards the center of the disc, jump up onto the label and perhaps even across it, and damage the stylus. On a manual turntable, the locked groove holds the stylus until someone lifts the tone arm off the record.

The lathe can be programmed via plug-in modules to produce any desired disc parameters, such as lead-in and end groove diameter; lead-in, spiral, and lead-out pitch; and cutter lift-off delay after reaching the end groove diameter. These parameters are achieved through control of the lead-in screw drive motor, in conjunction with the position sensing of the sled in the lathe bed and a solenoid which lifts and lowers the cutting head onto the disc.

Stereo Cutting Head

The stereo cutting head consists of a *stylus* which is mechanically connected to two *drive coils* and two *feedback coils* (mounted in a permanent magnetic field) and a *stylus heating coil* (wrapped around the tip of the stylus). This is illustrated in Fig. 14-11. When a signal is applied to the drive coils, the alternating current flowing through them creates a changing magnetic field which alternately attracts and repels the permanent magnet. Since the position of the permanent magnet is fixed, the coils move in proportion to the strength of the field created and the attached stylus moves with them. The drive coils are wound and mounted so that energizing either one alone causes the stylus to move in a plane that is 45° to the left or right of vertical, depending on which coil is energized. Feeding both coils an in-phase signal causes the stylus to move in the lateral plane, while feeding the coils an out-of-phase signal causes stylus motion in the vertical plane.

Fig. 14-11. Simplified drawing of a stereo cutter head.

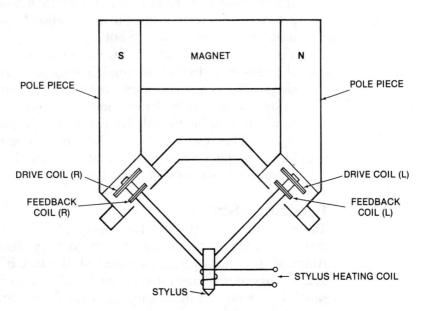

The feedback coils are attached to the shank of the stylus and are, therefore, moved whenever the stylus moves. The motion of these coils in the permanent magnetic field creates a current flow in them which is an accurate representation of stylus motion. By mixing the outputs of these coils out of phase with the input signal in a technique called *negative feedback*, several advantages are achieved:

1. The need for heavy mechanical damping materials to control the stylus motion is eliminated because the stylus responds more accurately to the drive signal. Without negative feedback, inertia can cause the stylus to overswing if it is not well damped. The lack of mechanical damping removes variations of damping characteristics due to age and heat from a hot stylus, which is characteristic of nonnegative-feedback cutting heads. In addition, negative feedback is more efficient because it controls the drive power while mechanical damping materials absorb it and dissipate it as heat.
2. Since the stylus follows the drive signal more closely, less distortion is produced.
3. The effect of irregularities in the hardness of the lacquer surface of the recording blank is reduced because the negative feedback senses them and changes the drive signal to compensate for them.
4. The signal-to-noise ratio of the disc is improved by approximately 16 dB at 10 kHz for a 6-inch groove diameter, producing an overall signal-to-noise ratio of about 70 dB when used in conjunction with a hot stylus.
5. The frequency response of the cutting head can be flattened or changed as desired by adding EQ to the negative feedback signal. The frequency range over which negative feedback can be used is limited by the mechanical resonance of the cutter head which shifts the relative phase of the input and feedback signals. This results in the feedback becoming positive at the higher frequencies. If the amount of feedback is not restricted at these frequencies, oscillations occur.

Recording Stylus

The *recording stylus* (Fig. 14-12) is made of sapphire because that substance is hard and can be ground to very accurate dimensions. Although it is not as hard as diamond, the lack of grain in sapphire and its ruggedness make it superior for disc recording purposes. The sapphire is mounted in an aluminum shank so that it can be attached to the cutting head. The *cutting face* of the stylus is ground flat and oriented so that the disc rotates into it. The tip of the cutting surface is ground to a 90° angle to form the *cutting edges*. The tip of the point is slightly rounded but must have a radius of less than 0.00025 inch. A *burnishing facet* ground into the stylus directly behind the cutting edge polishes the groove, as it is cut, to reduce noise. The dimensions of this facet are carefully controlled because excess width erases the high frequencies of the disc.

Fig. 14-12. A recording stylus.

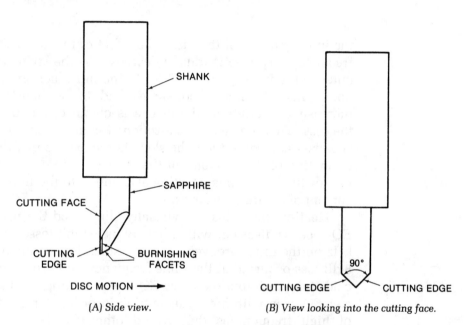

(A) Side view.　　　　　(B) View looking into the cutting face.

The *stylus heating coil* is a small coil of wire wrapped around the stylus tip through which current is passed to heat the stylus. Cutting with a hot stylus produces several improvements in performance. Since the linear distance per revolution travelled by the stylus decreases as it moves closer to the center of the disc, while the time required to complete each revolution remains the same, the *groove velocity* decreases as the *groove diameter* decreases. As illustrated in Fig. 14-13, groove velocity equals path length divided by the time needed to travel the path. The path length between lines A and B is less at groove diameter D1 than at diameter D2. Since the disc rotates at a constant angular velocity, the time it takes the stylus to travel an arc with angle θ is independent of groove diameter. As a result, the groove velocity at diameter D1 is lower than at diameter D2.

As the groove velocity decreases, more program material must be recorded per revolution and more information must be crowded on

Fig. 14-13.
Groove velocity
decreases as
groove diameter
decreases.

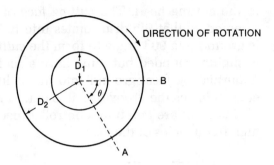

the inner grooves of the disc. The effect of this crowding is that high-frequency response gradually decreases as the stylus moves from the outer to the inner grooves. Prior to the introduction of cutting with a hot stylus, the high frequencies had to be boosted in gradually increasing amounts as the disc was cut in order to compensate for the loss. This *diameter equalization* was achieved by connecting the equalizer control to the lathe sled via pulleys, so that the EQ varied as a function of the diameter of the groove being cut. While compensating for the losses, this boosting also added to the noise and distortion content of the inner grooves.

Heating the stylus has virtually eliminated the need for diameter EQ because discs cut with a hot stylus exhibit losses of only 2 dB at 8 kHz on the inner grooves. This is a small loss as opposed to the 6- or 8-dB loss occurred at the same frequency and groove diameter with a cold stylus, without diameter EQ. Some cutting rooms still use diameter EQ even with hot-stylus cutting in order to recover the last 2 dB of high frequencies that would otherwise be lost on the inner grooves.

The smoothing effect of the hot stylus produces a signal-to-noise ratio which improves as groove diameter decreases. The signal-to-noise ratio of a disc cut with a cold stylus, on the other hand, worsens with decreasing groove diameter. Discs cut with a hot stylus are 2 dB quieter than discs cut with a cold stylus at the outer grooves and 18 dB quieter on the inner grooves. In addition, the smoothing action virtually eliminates groove modulation noise (similar to the tape modulation noise mentioned earlier). A hot stylus cuts through the lacquer coating of the recording disc much easier than a cold stylus, facilitating the cutting of lacquer discs that have hardened due to age. A cold stylus would produce inferior grooves, called *dry cuts*, on a hard disc. The stylus heat also eliminates the *horns* caused by the elasticity of the lacquer (Fig. 14-14). Horns are raised edges on the sides of the groove which are easily broken and can cause the groove walls to break or crack when the finished record is removed from the press, resulting in increased surface noise. If horns occur on a disc, they are removed by polishing the molds used on the presses. Since horns limit the level that can be cut on a disc, higher levels can be cut with a hot stylus.

Fig. 14-14. Horns resulting from cutting with a cold stylus.

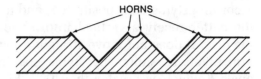

Stylus heat must be carefully set because too much heat can cause horns to form and the wrong amount can reduce the signal-to-noise ratio. Heater current is set by listening to the cutting head feedback outputs and adjusting the current for the minimum sputtering noise. Since groove velocity is constantly decreasing as the disc is cut, the heat applied per unit area progressively increases. As a result, lowest noise occurs only at the groove diameter at which the current was set. Ideally, heater current should decrease as the disc is cut. The signal-to-noise ratio of a lacquer master made with a hot stylus and a negative-feedback cutting head is about 70 dB, but the plating process used in the manufacturing of the finished records degrades this figure by adding ticks and pops to the signal.

The material removed from the lacquer disc by the stylus is called the *chip*. A tube aimed at the stylus and connected to the lathe vacuum system removes the chip as the groove is cut; this prevents the chip from blocking the path of the stylus. The suction is always started first, and the stylus heat is then applied as the cutting head is lowered onto the disc. If this is not done, the chip, which becomes very limp when stylus heat is used, may collect underneath the stylus before the suction can begin to act.

Distortion

Since the playback-stylus *tracking angle* has been standardized at 15° from vertical with the stylus pointing into the disc rotation, the cutting head must be angled so that it produces a groove that is oriented in the same way. This is complicated by the fact that the lacquer springs back somewhat after being cut. As a result, cutting heads are operated at a greater angle of about 18° in order to produce a 15° groove. This standardization of reproducer tracking angles is necessary because the nature of the recording stylus assembly causes vertical information to be cut in an arc, rather than strictly vertical. As the stylus moves in this arc, the groove angle becomes wider than 90°. If the playback-stylus tracking angle is not set to compensate for this increased groove angle, second-harmonic distortion will be introduced into the program.

As shown in Fig. 14-15A, with no vertical information, the recording stylus is in position 1, and a 90° groove angle is cut. When the cutter receives vertical information, the stylus cuts deeper into the disc. It does not move straight down, however; rather the support arm pivots as shown by stylus position 2. This produces a groove angle θ which is greater than 90°. Fig. 14-15B shows that since the playback-stylus vertical-tracking angle is standardized at 15°, the recording stylus cutting angle is set to ϕ (slightly greater than 15° to compensate for lacquer spring back) so that the groove walls will present a 90° angle to the playback stylus. Fig. 14-15C shows the vertical-tracking angle error with the stylus perpendicular to the disc. Peaks of the stylus motion occur before peaks of the groove motion, and dips in the stylus motion occur after dips in the groove.

Other potential sources of signal degradation, such as tracing distortion, the pinch effect, reaching the groove excursion limit, and mistracking can also occur during playback of the finished record. *Tracing distortion* results from the difference in the shape of the recording and playback styli. The recording stylus cutting edge comes to a sharp point while the playback stylus has a rounded point. Because of this, the playback stylus point of contact with the groove varies depending upon the instantaneous amplitude of the groove modulation, and the path traced in playback is not the same as that recorded (Fig. 14-16). The point of groove-wall contact for the playback stylus wanders as the modulation is traced, producing an output corresponding to the dotted path rather than the modulation. The distortion increases as signal level increases and as the wavelength of the recorded signal decreases and approaches the dimensions of the stylus tip. Thus, tracing distortion increases as the signal frequency increases and as groove diameter decreases, because both of these factors cause the wavelength of a signal to decrease (wavelength equals groove velocity divided by signal frequency, and groove velocity decreases as groove diameter decreases).

Tracing distortion refers to the inability of the playback stylus to follow groove modulations in the vertical plane. A similar problem in the lateral plane is called the *pinch effect*. Due to the triangular shape of the cutting stylus, the width of the groove measured perpendicular to the line cut by the stylus tip does not remain constant (Fig. 14-

**Fig. 14-15.
Tracing angles
for cutting and
playback styli.**

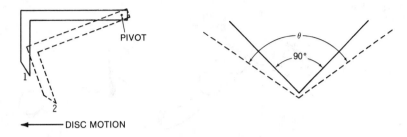

(A) Cutting at increased depth causes groove angle greater than 90°.

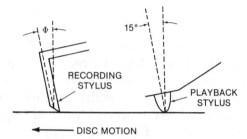

(B) Cutting angle is greater than playback angle.

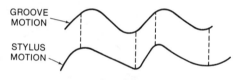

(C) Vertical tracking error.

**Fig. 14-16.
Tracing
distortion.**

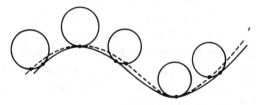

17A). The playback stylus is pinched where the groove narrows and, therefore, it rides higher in the groove. As the groove widens, the stylus rides lower in the groove (Fig. 14-17B). This motion adds a vertical component to the signal output that was not present in the input signal. The pinch effect is greater for high-level and high-frequency signals, for these cause abrupt changes in groove direction and, therefore, cause the width of the groove to become narrower.

The *groove excursion limit* is reached when the radius of curvature of the groove modulation is equal to the tip radius of the play-

Fig. 14-17. The pinch effect.

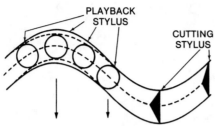

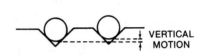

(A) Top view of a laterally modulated groove.   *(B) View in the plane of the groove, showing the playback stylus rising in the groove where the groove narrows.*

back stylus, preventing it from following all of the modulation. As shown in Fig. 14-18, the radius of curvature of the groove modulation is equal to the stylus tip radius, producing 50% second-harmonic distortion. Since the radius of curvature decreases both with an increase in level as well as with an increase in frequency, a limit exists on the level of high-frequency information that can be accurately reproduced.

Fig. 14-18. Side view of a groove illustrating the groove excursion limit.

All three of these types of distortion would be reduced if the playback stylus point was very small. Distortion would be eliminated altogether if the playback stylus was shaped like a recording stylus. A stylus with a small tip radius, however, would ride on the bottom of the groove, producing noise and other types of distortion, and a cutter-shaped stylus would tend to cut into the disc and erase high-frequency information.

A compromise solution is the use of a *bi-radial elliptical* stylus rather than the standard *conical* stylus. The conical stylus has a round tip with a radius of 0.6 mil. The elliptical stylus has an edge radius of 0.2 mil and a radius at right angles to the groove of 0.7 mil. The small-edge radius follows groove excursions more accurately than the conical stylus, while the larger radius prevents the stylus from hitting the bottom of the groove. Distortion is reduced somewhat, at the cost of increased wear on the disc.

A more effective way of reducing tracing, pinch effect, and groove-excursion limit distortion is through the use of a *tracing* simulator, such as Neumann's Model TS-66. All three of the discussed types of distortion are predominantly of the second-harmonic variety. Their effect can be reduced by *predistorting* the signal to be cut on the disc in an inverse manner to the distortion generated by the playback stylus (Fig. 14-19). The tracing simulator generates a voltage corresponding to the second harmonic of the program signal and adds it to the program out of phase with the distortion created in playback. This second-harmonic component is cancelled by the tracing, pinching, and groove-excursion limit distortion generated during playback, resulting in a disc with substantially less distortion and greater high-frequency level capability.

Fig. 14-19. Reducing playback distortion by predistorting the groove modulation.

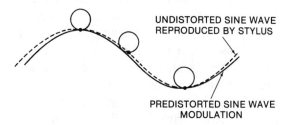

UNDISTORTED SINE WAVE
REPRODUCED BY STYLUS

PREDISTORTED SINE WAVE
MODULATION

Since the amount of distortion produced in playback is a function of groove velocity, which is in turn determined by the groove diameter for any particular turntable speed, a series of microswitches are connected to the lathe bed to sense the position of the cutter head. The microswitches enable the drive signal to be predistorted the proper amount for that groove diameter.

The tracing simulator can only optimize distortion for one tip radius at a time; it is usually adjusted for 0.6 mil, which is the international standard for the radius of a conical stylus. The elliptical stylus with its narrow side radius would not completely cancel the second-harmonic distortion added to the signal by the tracing simulator. Therefore, this type of stylus would not receive the full benefits of the distortion reduction.

Mistracking results from the loss of stylus contact with the groove walls during playback, due to recorded velocities in excess of that which the stylus can follow. The *recorded velocity* is measured in centimeters per second (cm/sec) and is determined by computing the distance that the playback stylus must move, laterally or vertically, per second from the unmodulated groove position to accurately track the groove modulation. Since recorded velocity is constantly changing, the value used is the instantaneous peak velocity.

Mistracking often sounds like a buzz or crackling on heavy modulated passages, or sibilance on vocals. Mistracking can also cause instruments which should have sharp clear sounds, such as bells, to produce a dull thud at the beginning of each note. Increasing the tracking force (pressure of the playback stylus on the groove) reduces mistracking. The maximum recorded velocity, listed by frequency, which can be accurately tracked at a certain tracking force is a function of the design and specifies the *trackability* of a stylus/cartridge assembly. The better the trackability, the lower the tracking force needed to prevent mistracking.

The pressure of the stylus results in some indentation of the groove modulations. As the radius of curvature of the recorded signal increases, the stylus indents the record surface more, resulting in less output. This *playback loss* becomes more severe as frequency rises and as groove diameter decreases, because both of these increase the radius of curvature. Typical finished records have playback losses of 3 to 4 dB at 15 kHz at minimum diameter, relative to the master tape. Stiffer record compounds are indented less and therefore have lower losses.

Most phono cartridges are designed using either magnetic or piezoelectric principles. The *magnetic cartridge* converts stylus motion into electrical current flow through the use of permanent magnets and coils. Designs are available using fixed magnets with moving coils, moving magnets with fixed coils, and fixed magnets with fixed coils in which the stylus motion varies the magnetic coupling to the coils. In all cases, the coils are angled 90° apart from each other and are oriented so that the signals in the outer wall of the groove produce current flow in the right-channel coil and signals on the inner wall produce current flow in the left-channel coil. The coils are phased so that lateral stylus motion produces in-phase outputs from both coils, while vertical stylus motion produces outputs that are out of phase.

The *piezoelectric* or *crystal* phono cartridge generates an output voltage proportional to the varying pressure applied to the crystal by the stylus motion. Crystal phono cartridges are much less expensive to manufacture than magnetic ones. They also produce higher output levels so that the extra stage of the preamplification required by magnetic cartridges is not needed. The magnetic cartridge, however, is quieter and operates at lower tracking forces, creating less wear on the grooves. Crystal cartridges are mainly used in low-priced record players.

Equalization

Disc recording uses pre- and post-equalization just as tape recording does, but the equalization is set to the RIAA (Recording Industry Association of America) standard rather than to the NAB standard (Fig. 14-20). This standard results in the level of bass frequencies being decreased and the level of high frequencies being boosted during cutting, with signals restored to flat by reciprocal equalization on playback.

Fig. 14-20. The RIAA disc playback equalization curve.

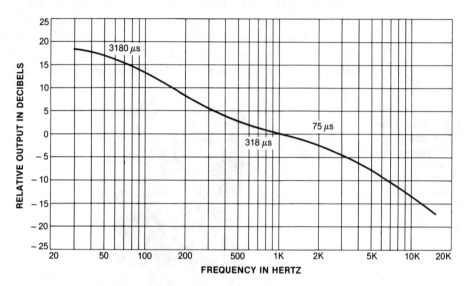

Two improvements in signal-to-noise ratio result from this. First, since most of the energy of music is in the low frequencies, the large groove excursions accompanying them would limit the maximum level cut on the disc, as well as limiting disc playing time. Rolling off the lows and restoring them in playback permits the upper portion of the frequency spectrum to be cut at a higher level. Second, since relatively little music energy is contained in the higher frequencies, these frequencies can be boosted without overmodulating the grooves. Reducing the high-frequency response to normal on playback also reduces the audibility of disc surface noise. The RIAA high-frequency EQ improves disc signal-to-noise ratio by about 8 dB.

The Mastering Console

The *disc mastering console* controls and monitors the signal cut on the disc and enables certain changes to be made. The Neumann SP-172 program controller is used as an illustration. Fig. 14-21 illustrates its equipment layout as follows: (A) Light-beam peak program meter, (B) Stereo VU meters (left and center) and correlation meter (right), (C) From left to right: channel gain controls, four Neumann equalizers for the program and preview channels, two high- and low-pass filters (each simultaneously controls the program and preview signals for its channel), the phase oscilloscope, the automatic banding unit, and the stereo master fader, (D) From left to right: the cutting controls, the console oscillator, and the metering and monitoring controls, (E) Patch bay, (F) Compressor/limiters for the program channels (stereo interconnected), (G) Four additional equalization channels to augment those in item (C) above, and (H) Dolby 301 unit. Items (F), (G), and (H) are not part of the SP-172 console but are interconnected with it in the particular installation shown in Fig. 14-21.

Fig. 14-21. The Neumann SP-172 Program Controller *(Courtesy The Master Cutting Room).*

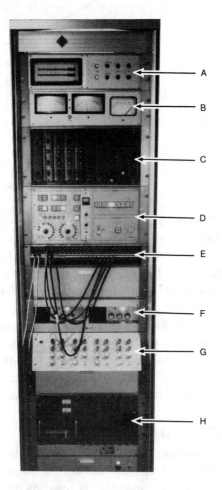

The console provides individual channel gain controls with a range of ± 10 dB in 0.5-dB steps to adjust the balance between channels as well as the relative levels of the different selections on a disc. A stereo master fader is provided to fade out any sounds not faded early enough on the master tape. A stepped *relative disc level* control adjusts the level of the signals cut on the disc relative to standard recording levels without affecting the console meter readings. Standard recording level is measured at 1 kHz and, for mono, is a peak stylus velocity of 7 cm/sec. For stereo, standard level is 3 dB lower per channel, at 5 cm/sec. This arises from the fact that a mono signal is cut laterally (Fig. 14-22) while the channels of a stereo groove are cut at 45° to the vertical. Since there is a 45° difference in the direction of cutting motion between a mono and a stereo signal, the lateral motion of a stereo groove equals that of a mono groove when stereo motion, in the plane 45° from vertical, is 0.707 times the mono motion (Fig. 14-23). This results in 0.707 times the mono output for each channel of the stereo groove. In decibels, this is a 3-dB less output per channel relative to the output of a mono groove.

Fig. 14-22. A mono groove producing lateral motion.

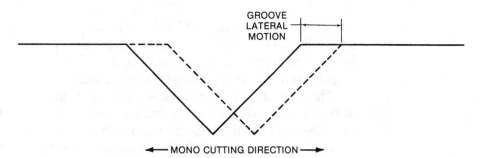

Thus, the stereo and mono standard levels produce the same lateral groove motion. The *relative disc level* control permits discs to be cut *hot* or *cold* with respect to standard level while still giving an on-scale reading on the meters. For example, if the record is to be cut 2 dB higher than standard level, this control is set for + 2 and the gain controls are adjusted so that the VU meters read 0 VU on the loudest sections of the program. These loud sections will then be cut 2 dB above the standard level.

The *preview offset* control is a stepped control which changes the gain only in the preview channels in order to adjust the sensitivity of the automatic pitch/depth control computer for optimum cutting of the disc. This control is usually set the same as the relative disc level control so that the computer will compensate pitch and depth for cut-

**Fig. 14-23.
Stereo groove
motion with one
channel
modulated,
producing the
same amount of
lateral motion as
the mono groove
shown in Fig.
14-22.**

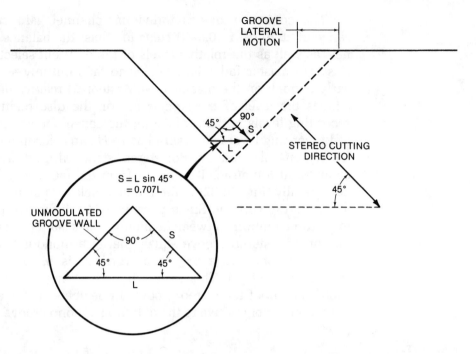

ting at other than standard level. It can also be set higher than the relative disc level control to cause extra expansion of the grooves on highly modulated passages and to reduce the chance of a highly modulated groove affecting a neighboring groove.

The console also provides equalizers to adjust the tonal balance of the program. Equalization can be used at this point to compensate for differences in the balance of selections mixed in different control rooms or mixed over different monitor speakers, or to compensate for deficiencies in the mixes themselves. In addition, EQ can be used to make the overall level of the program seem louder than it really is. Some record companies make a habit of boosting the midrange at about 5 kHz to achieve this effect. Four sets of equalizers are provided so that the same EQ that is used on the program channels can be added to the preview channels which feed the lathe computer. This is necessary to preserve the accuracy of the automatic pitch- and depth-control function.

The Neumann equalizers used in the SP-172 console provide EQ of ± 15 dB at 60 or 100 Hz and at 10 kHz in 11 steps; they provide EQ of ± 8 dB at 0.7, 1.0, 1.4, 2.0, 2.8, 4.0, and 5.6 kHz in 9 steps. The console also provides 12 dB/octave high- and low-pass filters with switchable cutoff frequencies of 8, 10, 12, or 14 kHz and 60, 125, 250, or 500 Hz, respectively.

External compressor/limiters can be connected in each channel to either increase the average level of the program or to prevent overmodulation. The compression mode reduces the dynamic range of the signal. The overall effect is that the soft passages of the signal seem louder than they would otherwise be, while the loud passages do not overmodulate the grooves. The limiter mode merely prevents overmodulation without reducing the apparent dynamic range. The compressor/limiters for each channel are interconnected for stereo use to prevent the center shifting that would occur if gain reduction was not the same in both channels. Ideally, compressor/limiters should also be provided for the preview channels, but this is not always done.

A high-frequency limiter is provided to reduce the very high stylus velocities created by high-level, high-frequency signals. High velocities are very difficult for the playback stylus to track without distortion or skipping. They can also cause the cutter head to overheat or blow its protective fuses if they are of long duration. The limiter causes gain reduction in proportion to the amount of high-frequency energy present, rather than in proportion to the overall level as with the compressor/limiter. It can also be used to reduce sibilance on the vocals in a mix, but its threshold must be set carefully so that cymbal crashes or other high-frequency information do not trigger it accidentally. The unit can be bypassed if no high-frequency limiting is desired.

Four types of visual monitoring devices are provided on the console. Two VU meters display average program levels for each channel corresponding to the loudness of the signal. Two light-beam display indicators show instantaneous program levels corresponding to maximum cutting stylus excursions. The third display is an oscilloscope in which the trace is driven horizontally by one channel and vertically by the other. The face of the scope is rotated 45° so that the horizontal and vertical axes are at 45° from true vertical, thus representing the shape of the record groove.

The oscilloscope represents the instantaneous relative phase of the two channels (Fig. 14-24). When no signal is present, the trace is a stationary dot at center screen. A left-channel signal produces a trace 45° to the left, while a right-channel signal produces a trace 45° to the right. When a stereo program is monitored, the trace produces swirls and other movements between these two positions. Complete separation between the channels, such as would result from out-of-phase signals, produces a vertical trace. As separation decreases in the two channels (i.e., as more information appears that is common to, and in-phase), the trace broadens. A mono signal produces a horizontal line. The movement of the trace thus represents the motion of the cutting stylus. The horizontal and vertical scope amplifiers are designed to compress the signals so that a full trace appears regardless of level differences between the channels. The mastering engineer watches the display and notes any sections of the program which have extreme separation. He must be sure that these vertical signals will not cause the groove to become too light, or skipping may occur on playback.

**Fig. 14-24.
Phase
oscilloscope
displays.**

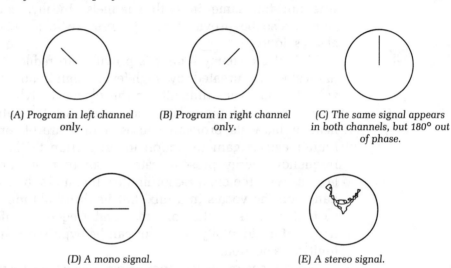

(A) Program in left channel only.

(B) Program in right channel only.

(C) The same signal appears in both channels, but 180° out of phase.

(D) A mono signal.

(E) A stereo signal.

Solutions to the problem of too much separation are to reduce the cutting level, to increase cutting depth, or to use an *elliptical equalizer* which reduces separation below its turnover frequency without affecting the high frequencies which provide most of the directional information. Decreasing the amount of vertical information also improves the mono/stereo compatibility of the disc. The Neumann elliptical equalizer has switchable turnover frequencies of 150 and 300 Hz.

The fourth indicator is the *correlation meter*. This is a center-zero scaled meter, similar in function to the phase oscilloscope except that it provides average rather than instantaneous phase information. Signal waveforms in the two channels, which are of the same polarity at a given point in time, are considered correlated, while those which are of opposite polarities are considered uncorrelated.

A positive reading on the meter indicates that the program has a predominance of correlated (in-phase) information and will produce more modulation in the lateral plane than in the vertical. A negative reading indicates that the program has a predominance of uncorrelated (out-of-phase or difference) information and will produce more vertical information than lateral. A zero reading indicates either equal amounts of correlated and uncorrelated information, or the lack of signal in one or both channels.

Identical signals in both channels (a mono program) produce a full-scale positive reading, while identical signals that are 180° out of phase produce a full-scale negative reading. A stereo signal with good separation causes a fluctuating reading which remains on the positive side of the scale. A negative reading indicates the possibility of shallow grooves and signal cancellations if the program is played in mono.

A *phase reversal* switch in the console compensates for the existence of a 180° out-of-phase condition between the channels by reversing the phase of one channel. An out-of-phase condition can arise due to miswiring of the console or the tape machine used to produce or play the master tape.

Other controls reverse the left and right channels and insert EQ and noise-reduction devices into the cutting chain. Another control provides a 14-dB gain increase for the meters, which is for use only in playback of test records for phono cartridge and cutter-head frequency-response calibration. (Test records are cut 14 dB below standard level to prevent overheating of the cutter-head coils by the sustained tones and to prevent excessive groove velocities at high frequencies.)

In addition, the console has a *mono mix* switch which combines the left and right channels in a one-to-one proportion to form a mono signal. This function is used when a mono single is to be cut and the producer does not feel it is necessary to make a separate mono mix of the song. Often the stereo mix used for the album becomes a single mix through the use of this function.

With the 45/45 cutting system, mono discs are produced without having to change the cutting head. This is accomplished by feeding the mono signal to both channels of the cutting head. When a separate mono mix is provided for cutting a single, the mastering engineer presses the *mono button* on the console. This button connects the left channel of the tape recorder to both channels of the mastering console.

There are several reasons for this: If the mono mix is recorded full track, it will be reproduced by the top gap on the half-track stereo playback head of the tape machine. If the mono mix is recorded in the half-track mono configuration, it is equivalent to the left channel of a half-track stereo recording, and it will also be reproduced properly. If the mono mix is recorded on both channels in the half-track stereo configuration, only the left channel is reproduced. This last case is important, for if the azimuth of the record head used to make the master tape is not set for optimum phase response, reproducing and mixing the two channels (either in the console or on a mono record player) could cause severe frequency cancellation. Playing back only the left channel of a mono tape also eliminates the necessity for changing the headblock on the tape machine when both stereo and mono discs are being cut.

One of the monitoring capabilities of the mastering console is the ability to switch between tape machine program and preview outputs, console output, several playback turntables, and the cutting-head feedback signal. The selected signal is displayed on the console meters as well as heard through the speakers. The cutting-head feedback signal enables the engineer to hear exactly what the stylus is cutting and, therefore, makes audible any imperfections or holes in the lacquer coating which might impair cutting.

Additional features of the Neumann console include a patch bay, a 5-frequency test oscillator, flashing illuminated push buttons (which signify nonstandard lathe functions, thus preventing the ruin of any lacquers), and remote controls for lathe and tape machines.

The actual mastering of a disc involves several steps. After the tape deck and Dolby units (if used) have been aligned to the level-set tones previously recorded at the head of the tape, the master tape is played and monitored on the speakers and meters. Any desired changes in tonal balance are made at this point, using the equalizers. If compression or limiting is desired, it is also done at this point.

The mastering engineer adjusts the individual channel gain controls so that the VU meters read 0 VU on the loudest portions of the program. He also watches the peak meters for excessive peaks and watches the oscilloscope and correlation meter for random or out-of-phase conditions.

If high peak-signal amplitudes or phase problems are noted, the engineer can make a *test cut* of that portion of the program. He examines the groove under a 156-power microscope to be sure that the peaks do not cause cutovers and that the random or out-of-phase signals do not cause the grooves to become too light. The test cut is also played to make sure that the signal neither distorts on playback nor causes the playback stylus to skip. Overmodulation can be eliminated by reducing the peaks with a limiter or by lowering the channel gain (or relative disc level). Inserting the elliptical equalizer or decreasing the gain (or relative disc level) will prevent the groove from becoming too light, as will advancing the preview offset; this will cause the depth of the cut to increase when random or out-of-phase signals are cut. The settings used are written on a card and filed in case additional masters or references need to be cut at a later date.

When an lp is being cut, the setup is slightly more involved. The EQ, compression, and level settings must be determined for each selection individually for optimal results. A decision must be made at this point as to whether each selection on the lp should be treated separately or as part of the whole. If separately, then each selection should be cut at the highest possible level, without introducing distortion or skipping, to produce the best signal-to-noise ratio and loudest playback level. If the selections are to be treated as part of a whole, the transitions between volume levels of adjacent selections should be smooth so that the lp flows easily from one to the next. The problem of transitions is the result of the different nature of each selection. For example, a voice accompanied by a single soft guitar will sound much louder than the same voice accompanied by a loud guitar, electric bass, piano, and drums, even though both mixes read 0 VU on the meters. In the latter case, the energy which produces the 0-VU reading is being created by five sources rather than two, and the proportion of energy provided by the voice is less than in the first case. If both selections are cut at the highest possible level, the voice will appear loud and large in size on the first selection and relatively small on the second one. The apparent change in size of the voice can be reduced by cutting the first selection at a lower, softer level or by compressing the second one so that it can be cut at a higher, louder level. By listening to the ending of one song and the beginning of the next, any necessary changes in levels can be made.

Mastering

The mastering engineer sets a basic pitch on the lathe. This pitch is determined by the duration of the program to be cut on that side of the disc and whether the music is loud continuously or has quiet sections as well. A lacquer is placed on the lathe, and compressed air is used to blow any accumulated dust off of the lacquer surface. Chip suction is started and a test cut is made on the outside of the disc to check groove depth and stylus heat. The start button is pressed, the lathe moves into the starting diameter, the cutting head is lowered onto the disc, the starting spiral and lead-in are cut as preset, and the tape machine is started automatically. As the side is cut, the engineer changes the console settings as previously determined. A photocell mounted on the tape deck senses the white leader tape between the selections on the master tape and signals the lathe to automatically expand the grooves to produce bands. After the last selection on the side, the lathe cuts the lead-out groove, and lifts the cutter head off the lacquer.

The master lacquer is never played because the pressure of the playback stylus would damage the recorded sound track. Reference lacquers, also called *reference acetates* or simply *acetates*, are cut to hear how the master lacquer sounds. Thus, any damage to the sound track is confined to the reference acetate, preventing any damage to the master lacquer from being transmitted to the finished record. The damage consists of high-frequency losses and increased noise and is aggravated by each successive playing. The damage done per play is proportional to the stylus pressure used. At a tracking force of 1 to 2 grams, most references remain usable for 8 to 10 plays.

Since the producer wants to hear any EQ, level, or dynamic range changes introduced in the mastering process, he orders a reference disc before ordering the master lacquers. After listening to the reference several times, on a system with which he is familiar, he can either approve it and order the masters, or order a new reference with different EQ, etc.

The producer does not have to be present in the cutting room when the references are cut, although some do prefer to attend so that they can hear the changes produced by the EQ and decide how they want the reference cut. Other producers either ship or bring their tape to the cutting room with instructions on how they want it to be cut, and order a reference disc. If the first reference is not acceptable, they order a second one specifying new EQ or level settings. This procedure is repeated until an acceptable reference is produced. The approved control settings are then used for the production of all the master lacquers.

The advantage of giving cutting instructions beforehand is that the producer does not have to make changes with an unfamiliar speaker system. Occasionally, a producer may not realize that there is a difference in sound quality between home speaker systems and cutting room speaker systems, due to the superior equipment and conditions available in the recording studio. In this case, the producer will be unaware that frequency balance heard in the cutting room speakers must be deliberately distorted in order to produce the desired sound effect on home speaker systems. His attempts to coordinate the two speaker systems will be continually frustrated, and he may be wise to forego attendance at cutting room sessions.

After the control settings are approved, the master lacquer can be cut at any time. Since the master lacquer must be plated within three days after it has been cut in order to prevent the groove walls from drying out, it is not cut until the record company tells the producer they are ready to press the disc.

The record company assigns each side of the disc a *master* or *matrix number* which the cutting room engineer scribes between the grooves of the ending spiral on the lacquer (Fig. 14-25). This number identifies the lacquer and any metal parts made from it, and eliminates the need to play the record to identify it. If a disc is remastered for any reason, some record companies retain the same master numbers, while others add a suffix to the new master to differentiate it from the previous one.

Fig. 14-25. The master number on a disc.

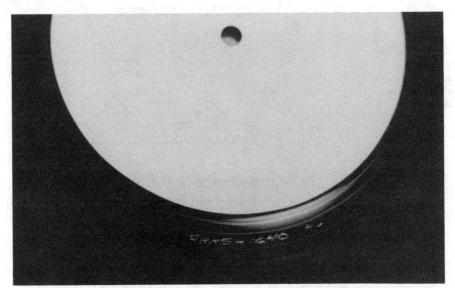

The engineer inspects the lacquers for any cutting defects and then ships them to the plating and pressing plants selected by the record company. Most record companies order three sets of masters so that the record can be pressed in three plants: one on the east coast, one in the midwest, and one on the west coast. Because each plant services the market closest to it, the cost of shipping finished records can be reduced.

While references are often recorded on both sides of a lacquer, masters are recorded only on one side of the disc. Cutting each side of a record on a separate lacquer allows the lacquers to be shipped with their blank sides touching and makes the production of molds for the record easier.

When the master arrives at the plating plant, it is washed to remove any dust particles and is then electroplated with nickel. When the electroplating is complete, the nickel plate is pulled away from the lacquer. This damages the master, so it can only be plated once. If something goes wrong at this point, the plating plant must order a new master from the cutting room.

The nickel plate pulled off the master is called the *matrix* and is a negative image of the master lacquer (Fig. 14-26). The bottom of the groove is the highest part of the matrix and the land is the lowest part. The excess nickel at the edges of the matrix is ground down and the matrix is attached to a metal or wood-backed plate to make it easier to handle. Since the groove walls containing the signal are now raised above the surface of the disc, they are very vulnerable and care must be taken to prevent damage to them.

Fig. 14-26. The matrix is a negative image of the master lacquer.

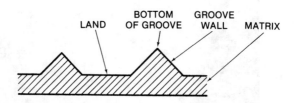

The negative image is electroplated to produce a nickel positive image, called a *mother*. Because the nickel is stronger than the lacquer disc, several mothers can be made from one matrix. Since the mother is a positive image, it can be played to test it for noise, skips, and other defects. If it is acceptable, the mother is electroplated several times, producing the *stampers*, which are the negative images of the disc that are used to press the records. The stampers are sometimes chrome plated to seal the nickel surface and make them last longer; however, the chrome tends to plate more on the groove edges than on the walls and increases the number of clicks and pops on the record. The plants that chrome their stampers feel it is better to have a few more clicks and pops than to have quieter records which deteriorate in signal quality as more records are pressed and the stampers wear out.

The plating process just described is called the *three-step process* to differentiate it from the *single-step process* which uses the matrix as the stamper. The single-step process is used when less than 200 records are to be pressed. The matrix, mothers, and stampers are called *parts*, and the plating process is referred to as *making parts*.

Stamping

The stampers for the two sides of the record are mounted on the top and bottom plates of a hydraulic press (Fig. 14-27). A lump of *vinylite* record compound, called a *biscuit* due to its shape, is placed in the press, sandwiched between the labels for the two sides. The press is closed and heated by steam to make the vinylite flow around the raised grooves of the stampers. Since the pressed record is too soft to handle when hot, cold water is circulated through the press to cool it before the pressure is released. When the press opens, the operator pulls the record off the mold. The pressing process causes excess compound to flow to the outer edges of the disc, so the disc is oversize. This excess, called *flash*, is trimmed off after the disc is removed from the press. The edge of the disc is then buffed smooth.

Record molds are designed to make the thickness of the finished disc greater in the label area and on the edge than in the groove area (Fig. 14-28). This is done so that when the discs are stacked after being pressed, their groove areas do not rest on each other. The raised outer edge, called the *groove guard*, must be the same thickness as the label area or *dish warp* will occur when the still soft discs are stacked. A disc with dish warp is saucer shaped rather than flat.

Other problems can occur at the pressing stage. *Pinch warp* can occur if the record cools too much before being removed from the mold, making it stick to the stamper. The operator can pinch the disc or even leave fingerprints on it in the process of pulling it off, causing warps in the disc.

Nonfill occurs when the vinylite does not flow around all of the grooves in the stamper. It is characterized by a swishing type of *once-around* noise; i.e., it occurs once each revolution of the disc. If the disc is held up to a reflecting light, the noise area becomes visible as a dull greyish area on the otherwise shiny surface. Nonfill can occur when insufficient steam is fed to the presses to heat them, so that the vinylite does not become hot enough to flow properly. Or, it can happen because the presses are running too fast for the vinylite to flow evenly throughout the entire mold.

The presses take more time to heat up when they are first started, so the first pressings are more likely to exhibit nonfill than the later ones. Occasionally, all the presses in a plant demand maximum steam at the same time. To satisfy this demand, a source of steam many times greater than that normally needed would be necessary. Since the cost of providing this extra steam is prohibitive, a few

Fig. 14-27. A record press.

Fig. 14-28. The profile of an lp.

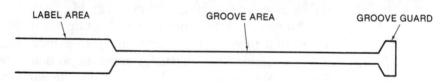

LABEL AREA   GROOVE AREA   GROOVE GUARD

cases of nonfill are unavoidable and have to be found and rejected by quality control.

The center holes of the discs are formed by the presses. Their location is determined by the mounting of the stampers on the press plates. The standard for 33⅓-rpm records is a 0.286-inch (+ 0.001, − 0.002-inch) diameter center hole, concentric with the recorded groove within 0.005 inch. For 45-rpm discs, the center hole is 1.504 (± 0.002) inches, concentric within 0.005 inch.

Since vinylite is actually a clear plastic material, carbon black is added to it as a *filler* in order to hide air holes and bubbles which would otherwise be visible and an unnecessary source of concern to the consumer. Some companies also add an antistatic compound to their vinylite to prevent the disc from attracting dust. This reduces surface noise resulting from dust in the grooves. While the antistatic compound is practical, it is expensive, and the amount added must be carefully controlled for too much will make the records noisy.

The hardness of the plastic has an effect on the frequency response of the disc because it affects the amount of playback loss. Stiffer compounds result in more high-frequency output due to less deformation of the groove modulations by the playback stylus. Since vinylite is stiffer than lacquer, the finished record sounds brighter than a reference disc cut with the same settings. The difference in frequency response is on the order of 1 to 3 dB at 10 kHz.

When a record is being readied for production, only a few discs are pressed initially. These are called *test pressings* and are used for checking the finished product for clicks, pops, skips, and nonconcentric center holes which can cause *wow*. The pressing plant checks several of these pressings first and, if they consider them acceptable, sends them to the record company which then sends one or two of them to the producer for his approval. Quality control by the producer at this point is very important; any defects not noted and corrected on the test pressing will occur on the finished product. Unacceptable records often slip by the pressing plant quality-control checkers, and record companies are sometimes slow in checking for defects. The producer and engineer responsible for the record are most familiar with the master tape and they are, therefore, best able to determine if any problems exist on the record. Since the presses do not have a chance to reach optimum temperature during the short run used for the test pressings, these records usually have somewhat more surface noise than the finished pressings.

The record companies assign each record a *release date*, at which time *disc jockey (DJ) copies* are mailed to radio stations. The *commercial* records are shipped to record distributors slightly after this date. The DJ copies of albums are the same as commercial copies except that the label attached to the disc usually contains timing information not included on the commercial copies. Also, the label is often a different color.

On DJ copies of singles, the "plug"side, or *A side*, is usually recorded in stereo on one side and in mono on the other side. Most commercial singles are now issued with a stereo A side and a different stereo B side. Although the consumer gets two songs, the record company doesn't have to worry about radio stations diluting the promotion for the plug side by playing the wrong song. Before stereo singles became popular, the plug side was pressed in mono on both sides of the DJ copies. The stereo side is intended for use by fm and a-m stereo pop music stations, while the mono side is for use by a-m mono stations.

Cassette Duplication

Over the past decade, the demand for pre-recorded music cassettes has risen from a back-seat market to one having equal billing with its vinyl disc counterpart.

It is evident from the previous section on disc cutting that a great deal of artistry and quality control goes into the manufacture of records. Contrary to public misconception, the same amount of care and quality control must be given cassette duplication in order to produce a quality product. Currently, there are basically three methods of cassette duplication: real-time duplication, bin-loop high-speed duplication, and in-cassette high-speed duplication.

Real-Time Duplication

Real-time duplication refers to the fact that the actual production process occurs in real time. That is, both the master playback machine and the slave recording machines are operating at the normal cassette speed of $1\frac{7}{8}$ ips. Thus, a program lasting 30 minutes would require 30 minutes of duplication time, and would produce as many programs as there are slave recording decks operating.

It is generally agreed that this format gives the highest quality of cassette reproduction, due to the fact that both the master and slave machines are operating at the standard cassette speed of $1\frac{7}{8}$ ips, a speed for which the medium is optimized. The audio signal is kept within the audio bandwidth and not shifted into a higher bandwidth as with the high-speed processes, thus allowing maximum signal, transport, and tape optimization. The recent marketing of dual-cassette tape recorders make use of this straightforward method and permits simple, high-quality, cassette duplication. Unfortunately, this method also permits the easy pirating of commercially mass-duplicated tapes, which both violates copyright laws and results in high-production and royalty losses.

The recorded tape, which is played on the master deck and from which the duplicated slave tapes are made, is called the *duplication master* or *dup master*. When using the real-time method, it is possible for the dup master to be in any audio format desired. It may take the form of a ½-track, 15- or 30-ips copy of the master tape, or it may be a ¼-track, 7½-ips copy, a cassette copy, or even the original master tape. Precise digital pcm copies of a master tape, with no tape generation loss, have also been used as dup masters by some quality-conscious duplication companies.

Often, the dup master may be recorded in ¼-track format onto ¼-inch, cassette, or, in some cases, ½-inch tape to ensure high quality. By recording in ¼-track format, the "side A" stereo program can be recorded in one direction and, then, by turning the tape over, a second, or side B, stereo program can be recorded onto the remaining tracks of the tape. Through the use of 4-track playback heads on the master machine and similar recording heads on the slave recorders, it is possible to record both sides simultaneously, in one single tape pass, thereby cutting the duplication time in half.

Bin-Loop High-Speed Duplication

The *bin-loop method* makes use of high-speed duplication in a process where the duplication takes place without the duplicated tape being housed in cassette shells. The tape is recorded onto reel-to-reel machines, which gives a higher quality and better tape handling at high speeds. In this method, a dup master is recorded from the master tape at 3¾ or 7½ ips, depending upon whether the program is voice or music quality, and, generally, in the ¼-track format previously mentioned. The dup master is then wound onto a bin-loop master reproducer (Fig. 14-29). This is a playback machine, which, instead of storing the tape on a reel, stores the tape in a special tape bin in a free-standing fashion. Once the tape has been wound into the bin, it is spliced end to end to create an endless tape loop. At this point, a recorded tone signal, corresponding to between 5–15 Hz at 1⅞ ips, is spliced between the program ends to mark the beginning and the end of the program loop on the duplicated slave programs. The program outputs of the bin-loop master may then be fed to any number of slave recording machines (Fig. 14-30). These slave machines are of the open-deck reel-to-reel type, with the exception that they are designed to record onto ⅛-inch cassette-grade tape, which is supplied on bulk 10½-inch pancake reels.

Fig. 14-29. The DP-80C Master Reproducer, a bin-loop transport machine *(Courtesy Otari Corp.).*

During the duplication process, both the master and slave machines operate at high rates of speed; at speeds of 60, 120, and up to 240 ips, with duplication ratios of 16:1, 32:1, and up to 64:1, respectively. Once the duplication process has started, the bin-master machine will play back the looped dup master repeatedly, at a precise high-speed ratio, onto the recording slave machines (Fig. 14-31). Thus, each successive repeated performance is recorded onto the bulk reels of cassette-grade tape.

With the duplication process occurring at ratios that are many times the normal speed, the frequency spectrum is also shifted up into a high-frequency range well beyond the audio spectrum. It is necessary that the head, the frequency response, and the bias currents be specially tailored for this demanding application. The bias current at these high speeds may commonly reach up to 3.5 MHz.

Fig. 14-30. A DP-1510 slave recorder showing the simplified tape threading path *(Courtesy Otari Corp.).*

Fig. 14-31. The bin-loop duplication system. A DP-80 Master Reproducer and 5 slave recorders *(Courtesy Otari Corp.).*

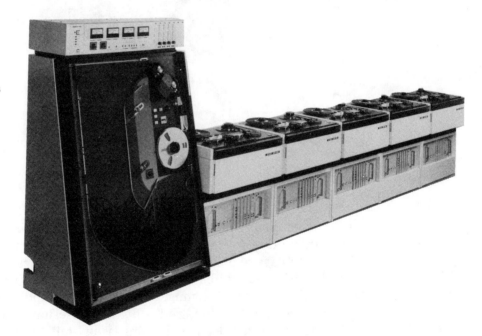

The next stage in the duplication process is to load the pre-recorded programs, which are now repeatedly recorded onto bulk tape, into cassette housings. This is done by a machine known as a self-feeding cassette loader (Figs. 14-32 and 14-33). The duplicated bulk tape is loaded into this device. A cassette-feed magazine is then filled with C-0 cassettes (cassettes loaded only with a short section of leader tape). Next, a C-0 cassette is dropped into the loading section and the recorded tape is automatically spliced onto the leader in the empty cassette, at the point where the beginning sensing tone of the program appears. The loader then fast forwards the tape, loading it into the cassette until the next tone is sensed, where it splices the program's end onto the cassette's tail leader and ejects the cassette. Then, the process repeats itself. Once loading has been accomplished, the final stage of the process is to label and package the cassette for sales and distribution. With a large-scale bin-loop method of production, it is possible to produce tens of thousands cassettes per day.

Fig. 14-32. The King Model 780 self-feeding cassette tape loader *(Courtesy King Instrument Corp.).*

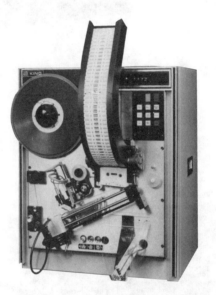

Product quality control is of great importance. The major emphasis in this process rests on the quality of the duplication master and the master/slave alignment. In the mastering process, it is most common for the master to be re-recorded into a stereo $\frac{1}{4}$-track, $7\frac{1}{2}$-ips format. In this format, care must be taken in the handling of distortion, high-frequency saturation, and a higher noise figure. Distortion and saturation are often dealt with using peak limiting and compression, and should be used sparingly. If alteration of the sound occurs and/or EQ changes are needed, the producer should be consulted as to these changes. The problem of noise may be improved with Dolby B noise-reduction procedures.

Fig. 14-33. The King Model 790 high-speed cassette loader *(Courtesy King Instrument Corp.).*

In-Cassette High-Speed Duplication

In-cassette high-speed duplication makes use of high-speed ratios (8:1 or 16:1) by reproducing from a dup master to a set of cassette slave recorders (Fig. 14-34) that are designed to handle the duplication of cassette tape which is already loaded into its shell.

In-cassette units are often self-contained, with both a master and one or two slaves being located in the same duplicating unit, with add-on slave machines often being available for later expansion. This allows for greater cost-effective design. The duplication master, with this method, most often takes the form of a stereo cassette, although units are available in a $\frac{1}{4}$-track reel-to-reel master format. This method is cost-effective but it may often have a trade-off of frequency, distortion, and wow/flutter limitations.

**Fig. 14-34.
Model DP-4050-
C2 8:1
in-cassette
sound
duplicator**
*(Courtesy Otari
Corp.).*

Compact Disc Mastering

The digital revolution, over the last three years, has made strong advances into the home audio market through the advent of the digital *compact disc* or *CD*. This silvery plastic disc, which is capable of reproducing over one hour of program material, has its information digitally encoded onto the reflective underside of the disc in the form of microscopic *pits*. When placed into a compact disc player, a *laser* is reflected off of this pitted surface and returned back to a *pickup* in the form of a digital stream of information. This digital stream of information is then restored back to music (sound) through complex digital-to-analog conversion procedures.

Figs. 14-35 through 14-42 show the steps followed in manufacturing a compact disc. The photographs are courtesy of the Digital Audio Disc Corporation, Terre Haute, IN. The photos and photo captions were originally used in the January, 1985 issue of *Digital Audio*, Peterbourgh, NH and are shown here with their permission.

The Mastering Process

Once the producer has finished a recording project, the first step is to master the musical material with the use of a Compact Disc mastering system. At the heart of one such system is the Sony PCM-1610 processor. The system is based upon a PCM sampling frequency of 44.1 kHz and it uses ¾-inch U-matic VCRs for music and information storage.

Fig. 14-35. Compact Disc manufacture begins at the editing stage with a specially encoded tape copy prepared in one of two identical control rooms. The sophisticated digital editing equipment can process music recorded on either digital or analog systems. *(Courtesy DADC and Digital Audio.)*

Fig. 14-36. Sony's Compact Disc Master Code Cutter produces a glass master by exposing a glass plate with a photo-resistant layer to a laser beam. Later, a CD player's laser will read the information pits cut into the layer and translate them into music. *(Courtesy DADC and Digital Audio.)*

Fig. 14-37. For the matrixing process, glass masters are placed into a high-speed, precision, rotary plating system designed by Sony. The plater produces metal masters and stampers. *(Courtesy DADC and Digital Audio.)*

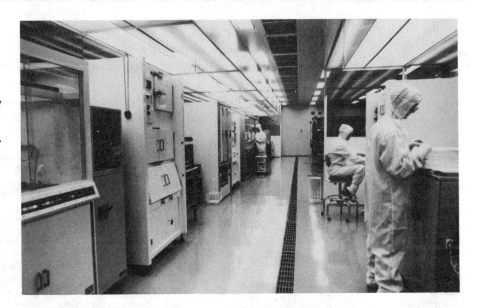

Fig. 14-38. After electro-plating, the stampers are placed into injection molding machines where they are injected with high-grade polycarbonate resin to create a transparent Compact Disc. *(Courtesy DADC and Digital Audio.)*

If a digital master is supplied that is already based on the 44.1-kHz sampling frequency, then the digital signal may be fed directly into the processor. Should the digital signal be based on another sampling clock rate, it will be necessary to translate from that sampling rate to the 44.1-kHz rate by way of a transcoder, such as the Studer

Fig. 14-39. Following the injection process, workers place a very thin layer of aluminum (the layer that reflects a Compact Disc player's laser) over the mold's pitted surface. An ultraviolet curing resin is applied then to form a protective coating. *(Courtesy DADC and Digital Audio.)*

Fig. 14-40. The plant's high-performance screen printer silkscreens individual labels onto each disc. *(Courtesy DADC and Digital Audio.)*

SFC-16. Analog masters must be digitally encoded and fed into the processor at this point.

Once the digital audio information has been fed into the system, the *subcode channels* may be programmed into the storage system. Subcodes are *event point codes* which are stored at the head of each

**Fig. 14-41.
Throughout the
CD
manufacturing
process, workers
in protective
clean-suits
check and
recheck the
discs with a
variety of
quality control
tests directed by
the CBS/Sony-
developed
computerized
automatic
inspection
system.**
*(Courtesy DADC
and Digital
Audio.)*

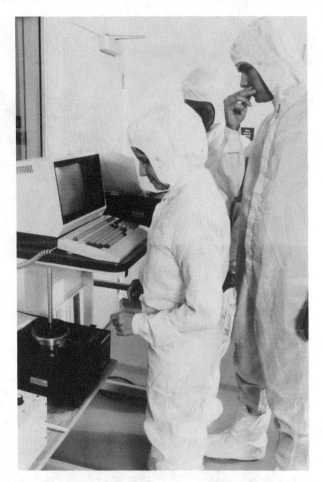

compact disc. It is the job of these codes to tell the CD player's micro-
processor just how many selections there are on the disc and where
the selections are located within the disc. At present, 8 subcode
channels are available on the CD format, although, at present, only
two (the P and Q subcodes) are used.

The encoder then splits the 16 bits of information into two 8-bit
words and applies error correction, through the use of the cross-
interleave Reed-Solomon code, in order to correct for lost or errone-
ous signals. The system then translates these 8-bit words into a 14-bit
word format for ease of recording onto disc.

After the encoder adds the subcode, breaks the samples into 8-bit
words, applies error correction, and converts the 8-bit words into 14-
bit words, the system begins constructing a methodic system for
compact disc operation known as a *data frame*. Each data frame
begins with one frame-synchronization pattern (27 bits) that tells the

**Fig. 14-42.
Stacked on
spindles,
finished CDs are
placed on
individual trays,
which then are
placed into
plastic
packaging cases.
Artwork and
any enclosures
to be included
are packed at
this time.**
*(Courtesy DADC
and Digital
Audio.)*

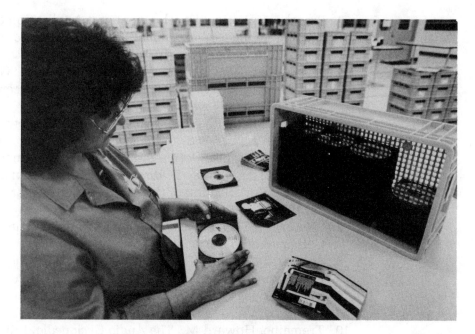

pickup beam where it is on the disc. Next is one word of subcode (17 bits), followed by twelve words of audio data (17 bits each), eight parity words (17 bits each), twelve more words of audio, and eight more parity words.

CD Cutting and Pressing

After the process described in the preceding section has been completed, the CD master is ready to be cut. This is done using a reusable glass master disc which has been specially prepared with a photo-resistant material. The master is cut using a 15-milliwatt laser. Once cut, the glass master is placed in a special developing machine which etches away all of the exposed areas, until it reaches the glass substrate. At this time, the CD manufacturing process becomes similar to the record pressing process. The glass master undergoes an electroplating process in which stampers are created from which the foil discs are pressed and finally encased in clear plastic for stability and protection.

References

1. Alexandrovich, George, "The Audio Engineer's Handbook," *db, The Sound Engineering Magazine*, Vol. 2, No. 4, pp. 6-8.
2. Birchall, Dr. Steven T., "CD: It's the Pits," *Digital Audio*, January 1985, pp. 26-30.

3. Braschoss, Dieter, "Development and Application of a New Tracing Simulator," *Journal of the Audio Engineering Society,* Vol. 19, No. 2, February 1971, pp. 108-114.

4. Gravereaux, Daniel W., and Benjamin B. Bauer, "Groove Echo in Lacquer Masters," *Journal of the Audio Engineering Society,* Vol. 19, No. 10, November 1971, pp. 847-850.

5. Hirsch, Julian, "Phono Cartridges," *Stereo Review,* Vol. 31, No. 1, July 1973, pp. 64-68.

6. Isom, W. Rex, Eric Porterfield, Stephen F. Temmer, Frank Gaudenzi, and Sidney Feldman; Panel discussion at meeting of the New York Section of the Audio Engineering Society, March 20, 1973.

7. "Neumann Tape-to-Disk Transfer System," *Technical Bulletin,* Gotham Audio Corp.

8. Schwartz, Arnold, "The Feedback Loop," *db, The Sound Engineering Magazine,* Vol. 4, No. 8, August 1970, pp. 8-12.

9. Silver, Sidney L., "Disc Mastering," *db, The Sound Engineering Magazine,* Vol. 1, No. 1, November 1967, pp. 7-9.

10. Tremaine, Howard M., *The Audio Cyclopedia,* Indianapolis: Howard W. Sams & Co., Inc., 1978.

11. Woram, John M., "The Sync Track," *db, The Sound Engineering Magazine,* Vol. 5, No. 2, February 1971, pp. 12-14.

12. Woram, John M., "The Sync Track," *db, The Sound Engineering Magazine,* Vol. 5, No. 3, March 1971, pp. 12-14.

13. *King Model 780 Technical Bulletin,* King Instrument Corp.

14. Birchall, Steve, "CD's Made in USA: High Tech Comes to Terre Haute," *Digital Audio,* January 1985, pp. 21-24.

15 STUDIO SESSION PROCEDURES

As was stated earlier, the first rule of recording is that there are no rules. This is only true insofar as inventiveness and freshness tend to play a key role in keeping our industry alive and new. However, in the recording process, there are guidelines and procedures to recording which, when followed, allow for a smooth, professional, recording session.

The multitrack recording process can be divided basically into four different studio procedures: recording, overdubbing, mixdown, and editing.

Recording

Before a recording session starts, the tape machines that are to be used should be cleaned, demagnetized, and then aligned for the type of tape being used for the session; preferably using the actual session tape. It is a good procedure to record a 100-Hz, 1-kHz, and a 10-kHz tone at 0 VU on all tracks at the beginning of the tape to indicate the proper operating level should realignment, a later overdub, or a mix at another studio be necessary.

It is often helpful to the musicians, producer, and engineer if all those involved in the recording process could sit down with the engineer, well in advance of the session, and discuss the instrumentation, studio layout, musical styles, and production techniques. This informs all parties of just what to expect during the session, as well as allowing those involved to get to know one another beforehand.

At the time of session, once the number and types of instruments being recorded are known, a pre-session setup should take place. Placement of instruments will often vary from one studio to the next because of the acoustics of the room, the number of instruments, the isolation between instruments (leakage), and the visual contact needed. Should isolation, beyond careful microphone placement, be needed, flats or baffles can be placed to prevent leakage of the louder instruments into the mikes designated to pick up the softer instruments. The setup should permit the musicians to see each other as much as possible so that they can give and receive visual cues. The arrangement of baffles and mikes will depend upon the type of sound that the producer wants. If the mikes are close to the instrument and the baffles are packed in close, a *tight* sound with good separation is achieved, while a looser, more live sound, as well as greater leakage, is achieved with the mikes and baffles farther away. An especially loud instrument can be isolated by putting it in an unused vocal or instrument booth. Electronic amplifiers played at high volumes can also be recorded in a vacant vocal booth, or can be isolated by building a box out of baffles to surround them and the mikes on all four sides and the top. Another approach would be to cover both the amplifier and the mike with a blanket or other flexible sound-absorbing material (Fig. 15-1), making sure that it does not interrupt the path between the amplifier and the mike. Separation can also be improved by placing the softer instruments in an isolation booth or by connecting the louder electronic instruments direct into the console, bypassing the miked amplifier option altogether. For a piano, leakage may be reduced by placing the mike inside it, putting the lid on its short support stick, and covering it with blankets (Fig. 15-2).

Fig. 15-1. Isolating an instrument amplifier by covering it with a sound-absorbing blanket.

Fig. 15-2. Preventing leakage from getting into a piano mike.

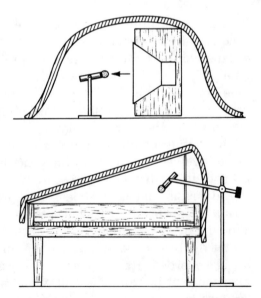

When using a directional mike, separation is often improved by putting a sound-absorbing baffle behind the performer or by placing him against an absorbative studio wall. Since the back of the mike is relatively insensitive, the leakage has to reach the live side of the mike in either a direct path or in the form of reflections from the hard surfaces behind the performer (Fig. 15-3). The directional mike should be angled so that any reflected sound from the instrument being miked, due to music stands or other nearby hard surfaces, reaches the dead side of the mike in order to prevent peaks and dips in frequency response that are due to phase cancellation between the direct and reflected sound.[1]

Fig. 15-3. A hard surface on the live side of a mike can put the undesired off-axis signal on-axis, by reflecting them.

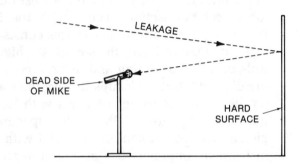

The microphones for each instrument are selected by either experience or by experimentation and are connected to the desired mike inputs. The input used for each mike should be noted on a piece of paper so that the console input module corresponding to each instrument can be easily found. Some engineers find it convenient to use the same mike input and tape track for the same instrument at every session. Thus, one engineer might consistently plug the bass guitar into input 1 and record it on track 1, so that he knows which fader controls the bass in both a record and a mix session without having to think about it.

Electronically amplified instruments with low-level, unbalanced, high-impedance outputs, such as guitars, can be recorded in the studio without using their amplifiers, through the use of a matching transformer or active matching amplifier. These devices convert the outputs to low-impedance, balanced signals which can feed a mike preamp. This is called picking up an instrument *direct*. Instruments are often picked up direct to avoid instrument leakage problems in the studio and to avoid noise and distortion resulting from the instrument's amplifier. Many newer electronic devices, such as electronic drum machines and synthesizers, which match more closely the impedance of studio equipment, may be plugged directly from the instrument into the line-level input of the console at the patch bay.

The instrument may then be played in the control room while listening to it over the main studio monitor speakers without fear of leakage. If studio direct pickups are preferred, both the direct box and the instrument amplifier can be fed from the instrument simultaneously, through the use of parallel jacks located at the direct box (Fig. 15-4). This enables the engineer to record the direct sound alone, the amplified sound picked up with a mike, or a combination of the two. The sound of a direct box is very clean, but more subject to string noise from instruments such as guitars than a miked amplifier sound. On most guitars, the lowest hum pickup and best tone for the direct connection occurs with the instrument volume control fully on. Since guitar tone controls consist of a variable treble roll-off, maximum control over the sound is achieved by leaving the tone controls on a treble setting and using a combination of console EQ and the different guitar pickups to vary the tone. If the treble is rolled off on the guitar, boosting the highs with EQ will increase noise that is picked up because of the high impedance of the guitar. A direct pickup connection can also be used with the speaker output jacks of instrument amplifiers if a means of attenuating the signal is used so that neither the transformer nor the mike preamp is overloaded.

Fig. 15-4.
Schematic for a
"direct box."

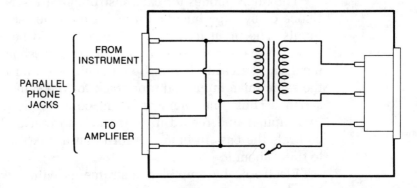

Over the last decade, drums have undergone a great deal of change with regard to playing technique, miking technique, and the choice of acoustic environment used for recording. The 1960s and 1970s saw the drum set commonly placed in a small isolation room called a *drum booth*. This booth effectively isolated the instrument acoustically from the rest of the studio and had the effect of tightening the drum sound because of the limited space (and, often, dead acoustics). It also isolated the musician from the studio causing a loss of direct-involvement feeling due to the physical separation. Recent years have seen the development of the electronic drum machine which eliminates the need for a physical drum set and studio mike placement, or the use of electronic contact drums which control a synthesizer and sequencer and allow conventional playing on placed keypads, which is then recorded directly from the synthesizer. The conventional drum set has also undergone changes in miking techniques due to the elimination of the drum booth, allowing the drums to be moved back into the studio with the other musicians. In many cases, as with new music production, drums are being distant miked, in addition to using conventional close-mike techniques, which gives a fuller larger-than-life sound to modern drum mixes.

If drums are to be recorded, the drummer should tune them while the mikes and baffles for the other instruments are being set up. Each drum head should be adjusted for the desired pitch and for constant tension around the rim by hitting the head at various points around its edge and adjusting the lugs for the same pitch around the head. After the drums are tuned, the engineer listens to each drum individually to make sure that there are no buzzes, rattles, or resonant after-rings when the heads are hit. Drums which sound great in live performance do not always sound that way when close miked. In a live performance, the rattles and rings are covered up by the other instruments and are lost before the sound reaches the listener. Close miking picks up the noises just as well as it picks up the desired sound.

If tuning the drums does not bring the extraneous noises under control, masking tape can be used around the edge of each head around the drum, as well as across the head, to dampen them out. Pieces of cloth, paper towel, or a wallet can also be taped to the head in various locations, determined by experimentation, to eliminate rings and buzzes. Although this method of head damping has been used extensively in the past, present methods use this damping technique with discretion and rely more fully upon proper tuning.

For studio recording, it is best to remove the entire built-in damping mechanism from the drum set, as it applies tension to only one spot on the head and, therefore, unbalances the head tension. The built-in dampers often vibrate when the head is hit and are one of the chief sources of rattles. The bass drum is damped by removing the front head and placing a blanket or some damping material inside it, pressing against the head by the beater. By adjusting the pressure of the blanket against the head, the bass tone can be varied from a resonant boom to a dull thud. Bass drums are usually recorded with their front heads removed, while other drums are recorded with their bottom head either on or off. Tuning the drums is more difficult if two heads are used because the tensions of the heads interact in producing the pitch, but a more resonant tone can be obtained than with only one head. After the drums are tuned, the mikes can be put in their desired positions, making sure that they do not get in the drummer's way. If the mikes are in the way, they may be hit by a stick or moved out of position during a performance.

Once the instruments and rough mike, pickup, and baffle placements have been made, headphones, equipped with enough extra cord to allow free movement, should be distributed to each player. The engineer then confers with the producer to find out how many instruments are to be used on the song, including overdubs, to determine how many tracks must be left open. This will influence the tracks to which the mikes are assigned, especially for the drums. If many instruments are to be recorded, the drums may be limited to two or three tracks. If there are plenty of spare tracks, five or more tracks may be used for the drums: bass or kick drum, snare, high-toms, low-toms, floor-toms, and, possibly, cymbal overheads.

When all the mikes have been set up, the engineer labels each input fader with the name of the corresponding instrument, either on a plastic write-in strip included above or below the fader on the console or on a piece of masking tape placed across the top or bottom of the faders. The mikes are assigned to the desired tracks, and the assignments are noted on a *track log* which is then attached to the tape box (Fig. 15-5). A rough headphones mix is set up so that the musicians can hear themselves, and the engineer asks them to play into the mikes one at a time. Starting with the EQ flat, the engineer listens for mike preamp overload and adjusts mike preamp gain, using the pad on the preamp, or on the mike if necessary, to eliminate any distortion. The EQ is adjusted to obtain the sound the producer wants on each instrument, and limiting is used if needed. If the desired sound cannot be achieved with a minimal amount of EQ, different mikes are tried until the sound is acceptable. The engineer and producer listen for any extraneous sounds, such as buzzes or hum from guitar amplifiers or squeaks from drum pedals, and try to eliminate them. This process of selective listening can be greatly helped by soloing the individual tracks as needed. If several mikes are to be mixed onto one track, the balance between them can be set at this point.

After this procedure has been followed for each mike and each instrument, the musicians should *run down*, or practice, the song so that the engineer and producer can listen to the instruments to hear how they sound together before they are recorded. All the drums are

Fig. 15-5. A track log for the instrument/ music track assignments *(Courtesy Steve Lawson Productions, Inc.).*

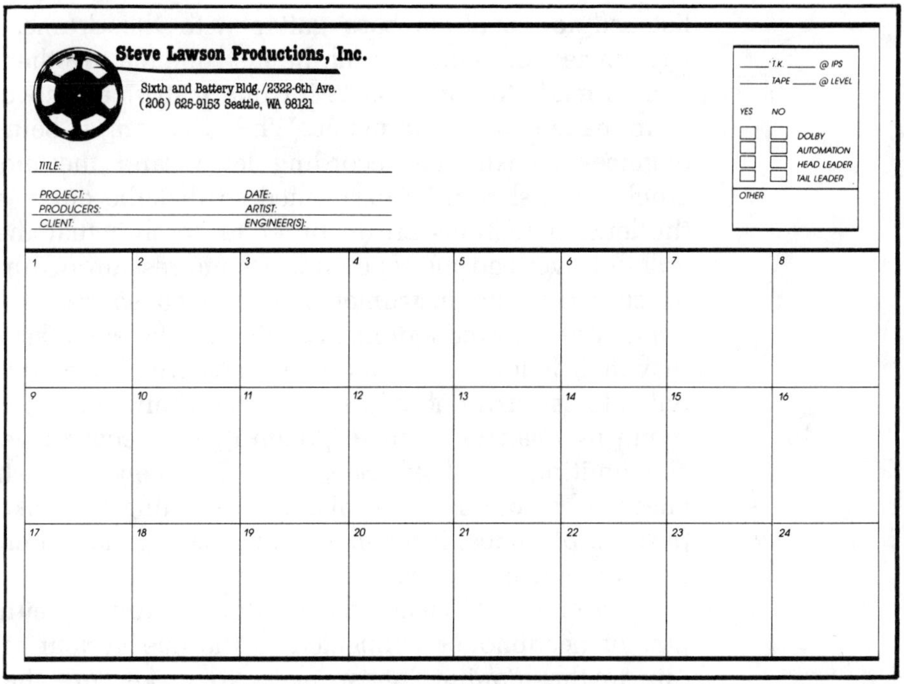

(A) Track identification chart.

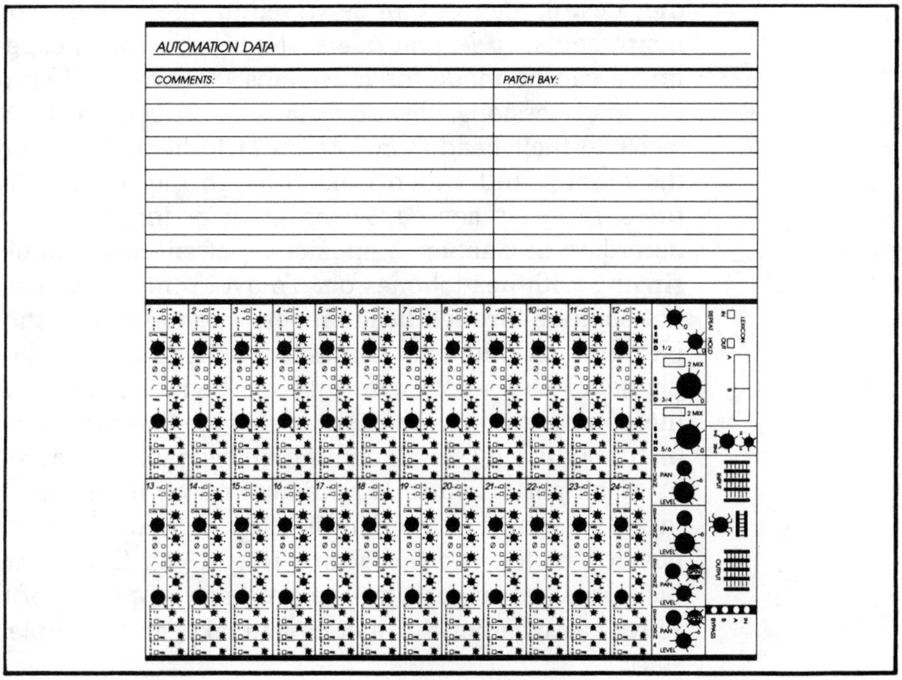

(B) Mixing and EQ group.

listened to, then the bass guitar with the drums, then the entire rhythm section, and then all the instruments together. Changes in EQ can be made to compensate for one instrument covering up another or to make them blend better. While the song is being run down, the engineer adjusts the recording levels and the monitor mix. The whole song should be performed so that the engineer knows where the loudest sections are in order to be sure that the recorded level will not overload the tape and, if compression or limiting is used, to be sure that the instruments do not cause more than the desired amount of gain reduction. Even though the engineer may ask them to play their loudest when the musicians are playing one at a time, each will almost invariably play louder when performing with others, requiring changes in mike preamp gain, record level, and compression/limiting threshold. Separation between the instruments can be checked by soloing each mike and listening for leakage. The relative position of mikes, instruments, and baffles can be changed, if necessary, to reduce leakage.

The engineer checks the headphone mix by either putting on a pair of headphones connected to the cue system or by routing the mix to the monitor loudspeakers to make sure that all the instruments can be clearly heard. If the musicians do not hear the sound they desire, the mix can be varied to intensify the sound of particular instruments. If several cue systems are available, several headphone mixes can be made for musicians that want different balances. During loud sessions, the musicians often require high sound-pressure levels in their headphones in order to hear the headphone mix above the room sound which leaks through the phones. Since high sound-pressure levels can cause the pitch of instruments to sound flat, as described in Chapter 2, musicians often have trouble tuning or even singing with headphones on. To avoid these problems, tuning should not be done through headphones. The musicians should be careful to play only as loud as is necessary to feel comfortable so that the headphone levels do not have to be too high. The same situation exists in the control room with respect to high monitor-speaker levels; some instruments may sound out of tune even when they are not.

Each performance is *slated* with both the name of the song and a *take number* for easy identification. A *take sheet* is kept to note the position of the take on a tape (Fig. 15-6). Comments are written on the take sheet to describe the producer's opinion of the performance, as well as whether it is a complete take, an incomplete take, or a false start.

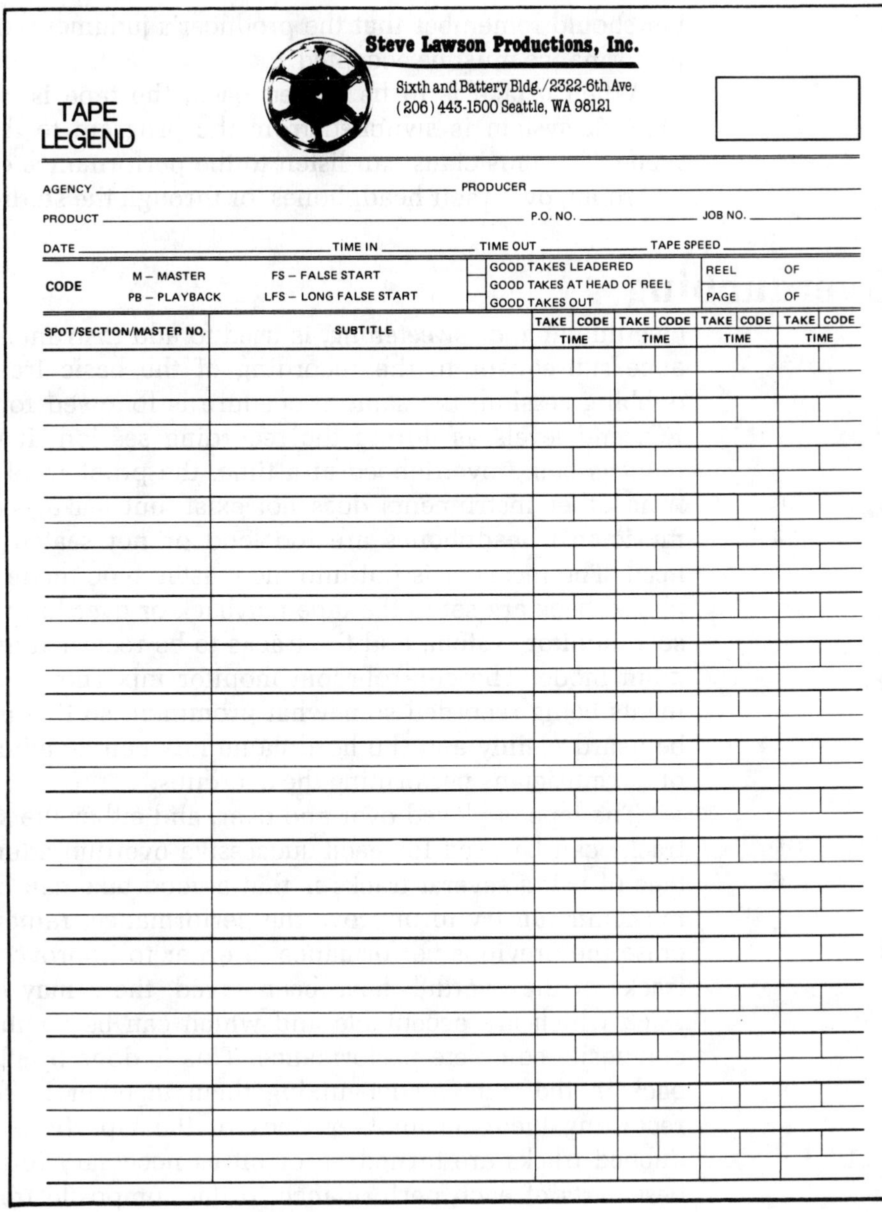

Fig. 15-6. Take sheet used for radio spots and for recording identification *(Courtesy Steve Lawson Productions, Inc.).*

Steve Lawson Productions, Inc.

TAPE LEGEND

Sixth and Battery Bldg./2322-6th Ave.
(206) 443-1500 Seattle, WA 98121

AGENCY _____ PRODUCER _____

PRODUCT _____ P.O. NO. _____ JOB NO. _____

DATE _____ TIME IN _____ TIME OUT _____ TAPE SPEED _____

CODE	M – MASTER	FS – FALSE START	GOOD TAKES LEADERED	REEL	OF
	PB – PLAYBACK	LFS – LONG FALSE START	GOOD TAKES AT HEAD OF REEL		
			GOOD TAKES OUT	PAGE	OF

SPOT/SECTION/MASTER NO.	SUBTITLE	TAKE	CODE	TAKE	CODE	TAKE	CODE	TAKE	CODE
		TIME		TIME		TIME		TIME	

During the recording, the engineer watches the level indicators and, if necessary, controls the faders to prevent overloading the tape. He also acts as another set of ears to listen to the performance. If the producer does not notice a mistake in the performance, the engineer might catch it and point it out. The engineer should try to be helpful, but should remember that the producer's judgment of the quality of a performance must be accepted.

When a take is to be played back, the tape is rewound and the monitor system is switched from the program to the tape playback mode. The musicians can listen to the performance either in the control room, over their headphones, or through the studio speakers.

Overdubbing

Overdubbing, or *sweetening*, is used to add instruments to a performance subsequent to the recording of the basic tracks. In an overdubbing session, the same procedure is followed for mike selection, EQ, and levels as during the recording session. If only one instrument is being overdubbed at a time, the problem of leakage directly from other instruments does not exist, but leakage can occur if the musician's headphones are too loud or not seated properly on his head. The recorder is put into the master sync mode, the tracks to be played back are set to the tape playback or overdub mode on the console monitor system, and the tracks to be recorded are set to the program mode. The control-room monitor mix should make the instruments being recorded somewhat prominent so that any mistakes can be heard plainly and the headphone mix can be adjusted to the taste of the musicians performing the overdubs.

The tape is played over and over, and either the same or different tracks can be used for each successive overdub attempt. The advantage of using several tracks is that a good take can be saved and the musician can try to improve the performance, rather than having to erase the previous performance in order to improve it. When several tracks of the overdub have been saved, there may be parts of each track which are acceptable and which can be combined to create a composite complete performance. This is done by playing the tracks back in the sync mode, mixing them together in the console, and recording them on another track of the tape (Fig. 15-7). The overdubbed tracks are turned on or off as necessary to transfer only the best parts of each performance to the composite track. Signals cannot be transferred to an adjacent track, however, because the crosstalk between the adjacent recording and reproducing heads would cause high-frequency oscillations (Fig. 15-8). This procedure, called *bouncing tracks* or *ping-ponging*, can also be used to mix entire performances onto one or more tracks, either to make the final mixdown easier or to open up their original tape tracks for additional overdubs.

**Fig. 15-7.
Bouncing
tracks.**

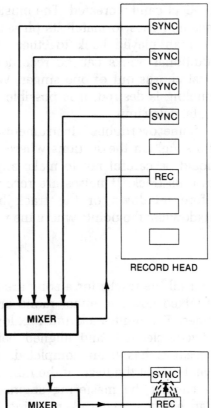

**Fig. 15-8.
Feedback path
when attempting
to bounce a
signal to an
adjacent track.**

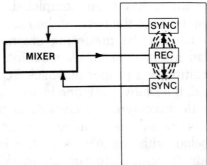

For a particularly difficult overdub, it may be easier for the musician to work on one part of the song over and over until it is performed properly before going on to the next section. This is done using a single tape track and *punching in* to the record mode at the start of the section to be worked on, *punching out* of record at the end of the section, rewinding the tape, and punching in again until the section is performed properly. The next section would then be recorded in the same manner until the entire song has been recorded.

Overdubbing is often used so that a musician can play or sing along with himself to make his performance sound fuller or so that an effect can be created. The musician listens to his original perform-ance and tries to match its phrasing as he overdubs. The two tracks are then played back together. This technique is called *doubling*. Additional tracks can be recorded in the same manner to make a vocal chorus out of one singer. When tracks are at a minimum and doubling is desired, it is possible to double electronically by using a digital delay unit.

Some overdubbed instruments only play during certain sections of a song. On the sections where he is not performing, the musician should be careful not to make any noises which would be recorded onto his track. If noises are recorded, they either have to be erased before mixdown or the track has to be shut down during the mixdown at the points where the noise occurs.

Mixdown

After all the tracks for a song are recorded, the multitrack tape must be mixed down to either mono or stereo for distribution to the con-sumer. The multitrack and mixdown tape machines must be demag-netized, cleaned, and aligned before starting. After the alignment procedure has been completed, the engineer should record 0-VU level tones at the head of the mixdown reel at 1 kHz, 10 kHz, and 100 Hz so that the mastering engineer (for whatever final format) can align his tape playback machine to play back these tones at 0 VU, resulting in a proper playback EQ for the tape. If the Dolby system is used, the Dolby level tone is used in addition to these tones.

In mixdown, the console is placed into the mix mode (or each input module is switched to the line position) and the faders are labeled with the names of the instruments they control (either on their write-in strips or on masking tape). The engineer sets up a rough mix of the song by adjusting the levels and the left-right pan positioning in the way that he thinks the producer would want them. The producer listens to this mix and asks the engineer to make spe-cific changes, such as "make the guitar louder" or "add some bass to the voice." The instruments are often soloed one by one or in groups, and EQ changes are made. The engineer tries to translate the pro-ducer's rough descriptions of sound character, such as "fat," "thin," "round," or "wooden," into settings on the equipment and console. Compression and limiting are used on the individual instruments as needed, either to make them sound fuller and more consistent in level or to prevent them from overloading the mixdown tape when they are raised to the desired level in the mix. At this point, should it be necessary, the console's automation may be brought into place. Once the mix begins to take shape, echo and effects are added to give close-miked sounds a more live spacious feeling and to help the instruments blend.

If the mix is not automation assisted and the fader settings have to be changed during the mix, the engineer marks the different levels on the fader scale with a grease pencil and learns when to move each fader from one mark to another. If there are more changes needed than he can handle alone, the producer or artist might help by controlling certain faders. It is best, however, if the producer does not have to handle any controls because then he can concentrate fully on the music rather than the mechanics of the mix. The engineer listens to the mix from a technical standpoint to detect any sounds or noises that should not be present in the mix. If noises are recorded on tracks not in use during a section of a song, these tracks can be shut off until needed. After running down the song enough to determine and learn all the changes, the mix can be recorded and the ending faded out with the master or submaster faders. Songs with exceptionally difficult control changes can be mixed in sections which can be spliced together afterwards.

The different takes of a mix should be slated as they are recorded, and a take sheet should be kept noting the differences between takes. When an acceptable take is recorded, it is leadered at the tail end (and, preferably, at the head as well) so that it can be easily found. The leader can be inserted either roughly or tightly depending upon whether the producer is in a hurry to go onto the next mix.

The beginning of a mix is listened to at high volume and the tape is moved back and forth over the heads by hand, with the tape machine in the *stop-edit* mode (head lifters defeated), to a point just before the first sound of the performance begins. The tape over the playback head gap at this point is marked with a grease pencil. If there is no noise directly in front of this spot, it is good practice to cut the tape a half inch before the grease pencil mark as a safety precaution against editing out part of the first sound. If there is noise ahead of the first sound, the tape should be cut at the mark and the leader inserted.

The tail of the song must be monitored at even higher volume because it is usually a fade-out or the overhang of the last note and is therefore much softer than the beginning of the song. The tape is marked and cut just after the last sound dies out to eliminate any low-level pops that may have been recorded when the bias fed to the record head was shut off and, also, to get rid of the tape hiss from the blank tape.

Mixes should be made at consistent listening levels because the variation in the frequency response of the ear at different sound-pressure levels results in a mix sounding quite different at different monitoring levels. The level used should ideally be the same as that at which the listener will hear the record. Since most people listen to music at moderate volume, moderate monitoring levels (80- to 90-dB spl) should be used.

The mix should be tested for mono/stereo compatibility to see what changes in instrumental balances will occur when the material is played in these different formats. If the changes are drastic, the original mix may have to be modified to make, for example, an acceptable mono video mix of a stereo LP mix. A mono mix should also be monitored over a car radio type of monitor, such as the Auratone™ speakers to see how it will sound on a system with limited frequency response. If the mix is for a single rather than for an LP, the entire mixdown session may be done at low volume over the car radio speakers.

Editing

After all the mixes for an LP are completed, a master-mix reel is made up for each side of the record. The producer and artist decide upon the sequence of the songs on each side on the basis of their tempos, their keys, which songs seem to flow best into one another, and which songs will attract the listener's attention best. The engineer edits the mixes out of their original reels and splices them together in sequence on their master reels, tightening up the leaders at the same time if this was not already done. The level-set tones are included as the first band on the master reel for side 1 of the record.

The length of time between the end of one song and the beginning of the next can either be a constant 3 to 6 seconds, or an amount that varies with the relevance of one song to another. Decreasing the time between songs can make a song seem to be a continuation of the previous one if they are similar in mood, or can make a sharp contrast with the preceding song if the moods are dissimilar. Longer times between songs let the listener get out of the mood of the previous song and prepares him to hear something that may be different, without accenting the contrast between them. If a crossfade is desired at this point rather than silence between the songs, it may be done as described under *backward* tape recorder effects in Chapter 8.

The length of leader tape used determines the time between songs. Paper, rather than plastic, leader is used because the plastic can cause static electricity pops. Blank tape can be used rather than leader, but it does not provide a visual division of the songs on the reel, making it more difficult to find a particular song. It also produces tape hiss between the songs, rather than silence.

When the sequencing is complete, several cassette or 7½-ips, ¼-track dubs of the master reels are made for the producer, the artist, and the record company executives, so that they can hear the mixes and approve them before the record is pressed. A *safety copy* of the master mix tapes is made at 15 ips, 30 ips, or the recorded digital format before the master mixes leave the studio. This copy serves as back-up protection in case the original mixes are lost or damaged. Although it is one generation removed from the master, and has 3 dB more tape noise, it can be used to cut the lacquer masters if necessary.

References

1. Malo, Ron, "Phase and the Single Microphone," *Recording engineer/producer*, Vol. 2, No. 6, November/December 1971, pp. 17-18, 31.
2. Woram, John M., "The Sync Track," *db, The Sound Engineering Magazine*, Vol. 6, No. 5, May 1972, pp. 10-12.

16 TOMORROW'S INDUSTRY

With the fast-paced changes in technology that have been brought about mostly with advent of the integrated circuit, it is difficult not to ask the question: "Where is the industry going?" For anyone to answer that question definitively is to invite being totally on the wrong track (especially to those of you who may be reading this in the far future). However, there are many definite trends which are just now beginning to show all around us, both in the studio and in the control room.

In looking to future industry trends, one needs to look at both sides of the coin: the artistic movement and the movement of technology. The two often move hand in hand, but in the production work of the 1980s, there are some interesting diversions. Artistically, the recording studio is beginning to enter into a renaissance of production and engineering. For the first time, the industry is beginning to see a combining of present-state music-production equipment and techniques with those which have been used in the past.

Just as vintage clothing eventually comes back into style, so has the era of the larger studio design, the distant miking practices, and the revival of the ribbon and tube-condenser microphones which have long been relegated to the limbo of the mike closet, as well as the "firing up" of the older recording gear for its special "tube" sound. What is happening, for the first time, is that the industry, generally as a whole, is saying that we can actually learn something from the past and maybe, just maybe, they may have had some really good things going. There have always been a precious few who have always subscribed to this notion, but now the whole industry has begun to turn its head.

Of all the past styles being revived, one of the most impressive is the return of the larger studio acoustic design with its greater cubic footage, higher ceilings, and live reflective room character. This design style was prominent during the forties and fifties when the majority of the recording studios were owned by the major record companies. In the sixties, with the advent of multitrack recording, this gave way to smaller absorbative studios, which effectively eliminated studio acoustics and caused a great deal of reliance on overdubbing due to the small space and increased isolation. Due to the revival of the larger sound studio, production styles in the 1980s, more than ever before, are calling for live sound and the utilization of distant miking techniques.

Distant miking, in many situations, permits a very different sound-character pickup than what we may be used to in modern music releases, thus often offering a new eighties sound. For example, the distant miking of a drum set has added a new *larger-than-life* sound that is now popular on many current pop music mixes. For many acoustic instruments, a distant pickup yields a greatly improved sound, allowing the entire instrument to be picked up instead of just a small portion, giving a natural overall sound quality. Stringed instruments, where the entire body of the instrument is the acoustic medium, often come under this category.

In taking from the past that which has withstood the test of time and combining it with the best of present-day practices, the recording industry is learning a very valuable lesson, which results in a growing repertoire of available production styles, not just the current latest trend. By contrast, modern recording technology is bringing the future into the present at a faster pace than at any time since the development of electrical recording. This has been brought about mainly by the development of the integrated circuit and a growing computer technology. The obvious buzzword for the rest of the century, a direct result of these developments, is *DIGITAL*.

A little over a decade ago, Lexicon Inc. created the Delta T digital delay unit. Viewed as a novelty, it was a marvelous, yet mere shadow, view of what was yet to come. The early 1980s began to see the production of digital effects equipment, such as delays, pitch changers, and early reverb devices. Digital recording was just coming onto the recording scene with the early PCM rotary-head recorders and the fixed-head multitrack recorders; the latter, however, required a great deal of care and maintenance in its early stages. We were all saying that digital processes would someday affect our lives, but believed that it would be many years before it would make any real difference. Fast-paced advances in electronic and computer technology have compressed these "many years" virtually into *today*. Digital recording, on some level, is now within the reach of the average studio budget, with the use of the PCM/video recorder combination, allowing live, stereo, digital recording or mastering directly from a multitrack tape. The digital audio stationary head, with track formats ranging from 2 to 32, is becoming a necessary and attainable low-maintenance item, although it is still an expensive initial outlay.

Another wave of the future is just beginning to show itself with the development of digital signal processing. The present-day control room is filled with many separate signal-processing devices, such as equalizers, effects processors, and reverberation units, to name but a few, with each device performing its own separate function. As signal processing expands, it is becoming the trend to design digital processors which can perform a multitude of functions, simply by reprogramming the device. This new age of multifunction computer-based systems has been best described by David Schwartz of Compu-Sonics as being a "chameleon—capable of changing its function at will." Thus, one device may be used as a digital delay, a reverb unit, a flanger, a phaser, or any number of effects units which have not even been invented yet. The changing from one program to another may take any form, from the push of a button recalling a preset program to a custom program that is tailored to the engineer or producer, and which may be stored on a cassette, a floppy disk, or a master tape in an encoded form.

It seems very likely that, in the future, these devices shall be able to perform any number of functions simultaneously, with just a simple set of adjustments needed to control the selected program's parameters (with total setup recall). This may permit the control room of the future to house some 1- or 2-effects devices, replacing the present multitude of devices. Even more likely is the possibility of accessing and storing various programs directly from the console's own computer, thereby eliminating unneeded outboard equipment altogether.

The marriage of the digital age and the multitrack console has already begun to become quite popular. Digital signal routing within the console allows the assignment of track selections, sends, and returns, as well as a multitude of signal-switching functions. This helps to eliminate the use of expensive and potentially defective switches throughout much of the signal path. Digital routing also permits the selected signal paths to be digitally encoded and stored for later recall. This, in combination with automated mixdown, gives a great deal of recall flexibility.

As we learned in Chapter 9, Neve Incorporated has moved multitrack recording a step further with the introduction of the DSP console. Designed as a totally digital console, all of its signal routing, as well as the signal path itself, is digitally encoded. Since all signal routing, EQ, sends, returns, panning, and signal processing, plus virtually all the console functions are digitally stored in the DSP's computer, it is therefore possible to double up on the use of the effective controls (Fig. 16-1). This permits the same set of controls to, for example, digitally set the levels for effects send number 3 and, then, be reprogrammed to set the foldback level, simply by addressing the computer and thus changing the control's function. Besides the potentially increased purity of sound, due to the lack of excessive analog-to-digital conversions, there is the added advantage of having absolute total console-setting recall, allowing the console to be set up to run a previous remix session in just seconds. As digital technology becomes more available and cost-effective, it is very likely that this design form will become more accepted and more what the computer people call "user friendly."

Fig. 16-1. DSP input and aux control panel *(Courtesy Rupert Neve Inc.).*

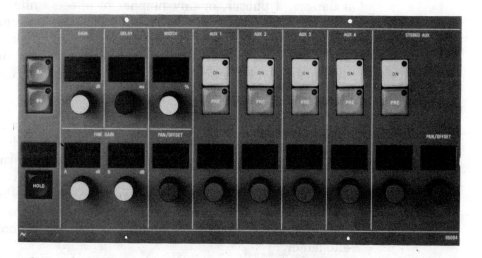

Recording gear is not the only facet to be affected by digital procedures; the world of music in recent years has been revolutionized by the digital wave—through the introduction of digitally controlled synthesizers, emulators, and sequencers. After having survived many early years of an identity crisis, the synthesizer has emerged as an established tool, both on stage and in the studio. Digital waveform creation and recreation has made it an instrument capable of producing an immensely wide variation of sound characteristics, as well as the close approximation of many acoustic instruments. Many of the new computer-controlled keyboard devices, designed to emulate established musical instruments, are giving artists and producers a great deal of flexibility, allowing them to compose or put together a complete demo or finished project themselves with simply one keyboard instrument.

Many synthesizers and sequencers have the ability to reproduce a number of short, digitally recorded, audio segments at will from their integrated circuits; segments which can be played back either at will or sequentially. This is the case with the electronic drum machine, which is now commonly heard on many popular releases. A step further is the ability to record any sound digitally into a new form of synthesizer, which can transpose the recorded segment into all the notes of the keyboard polyphonically. This breed of instruments, although relatively new, can recreate many instruments, including drums, quite realistically.

Often such synthesizers have the ability to record musical passages, precisely as played, into a digital memory "track." Some of these units have the ability to record up to 32 tracks, making it possible to record an entire composition digitally without the need of recording equipment. When used in conjunction with a musical instrument digital interface (MIDI), it is possible to sync any number of electronic musical devices to perform a musical piece. With SMPTE/MIDI, it would be easy to synchronize such an electronically recorded performance to a tape track for added vocals.

What does all this tell us technologically about the trends of the industry? It says that throughout all of these developments, the signal is *digital* and they all speak the same language, *digital words*. Not only are these devices potentially digitally compatible with each other but, in the future, all of the functions of these devices may be housed in a single "digital chameleon" device.

One such recording device, out of the several existing systems which employ computer-based random access audio technology, is the Synclavier® Digital Audio System from New England Digital. The basic central synthesizer/keyboard system allows for sounds to be digitally sampled directly into the system's memory, where they may be further processed by way of 16-bit resynthesis or FM synthesis. These sounds may then be directly filed to hard disk for recall at a later time.

The Synclavier® Digital Audio System and the Direct-to-Disk Digital Multitrack Recorder® combine to form a complete computerized recording environment for effects, music, and dialogue postproduction (Fig. 16-2). The heart of The Tapeless Studio is the high speed Synclavier computer which integrates with the direct disk processor to provide for the storage of digitized audio to multiple hard disks (for configuration within 4, 8, 16, and 32 track production) and/or to optical disks. The latter is able to offer a full 2 gigabytes of mass memory for on-line storage of sound data.

Optional system interfaces include a digital guitar, MIDI, SMPTE, multichannel analog output matrix, digital I/O modules and external timing modules. These combine to create a powerful music/ multitrack recording system for use within live performance, music production, and video/film postproduction.

Fig. 16-2. The Synclavier® digital audio system and Direct-to-Disk® multitrack recorder *(Courtesy New England Digital).*

References

1. *Synclavier® Digital Audio System and Direct-to-Disk™ Multitrack Recorder* brochure, New England Digital.
2. *Neve DSP Console Product Bulletin*, Rupert Neve, Inc.

INDEX

MORE
FROM
SAMS

☐ Crash Course in Electronic Technology

This book uses a step-by-step format to introduce electricity and moves quickly into circuit basics and electronic devices. Hobbyists, technicians, students and laypersons will learn the basics of electronic communications, controls, motors, test equipment, and troubleshooting—all presented in a self-paced, self-instructional format and loaded with clear illustrations and examples. Louis E. Frenzel, Jr.
ISBN: 0-672-22494-1 . . . . $18.95

☐ Handbook for Sound Engineers: The New Audio Cyclopedia

This up-to-date audio reference guide helps you make everything on tape or disc crystal clear. Covers all aspects of audio engineering, including tools for digital and disc recording, sound generation, sound system design, acoustics, and more. Concise explanations of terminology and illustrations are included.
Glen Ballou, Editor.
ISBN: 0-672-21983-X $79.95

☐ Reference Data for Engineers: Radio, Electronics, Computer, and Communications (7th Edition)

This book presents information essential to engineers, covering such topics as: digital, analog, and optical communications; lasers; logic design; computer organization and programming, and computer communications networks. An indispensable reference tool for all technical professionals. Edward C. Jordan, Editor-in-Chief.
ISBN: 0-672-21563-2 . . . . $69.95

☐ Introduction to Professional Recording Techniques

This all-inclusive introduction to the equipment and techniques for state-of-the-art recording delivers a comprehensive discussion of recording engineering and production techniques, including special coverage of microphones and microphone techniques, sampling, sequencing, and MIDI. Provides a wealth of valuable information on topics not found in other recording titles. Bruce Bartlett (John Woram Audio Series).
ISBN: 0-672-22574-3 . . . . $18.95

☐ Principles of Digital Audio

Here's the one source that covers the entire spectrum of audio technology. Starting with the fundamentals of numbers, sampling, and quantizing, you'll get a look at a complete audio digitization system and its components. Gives a concise overview of storage mediums, digital data processing, digital/audio conversion, output filtering, and the compact disk. Ken C. Pohlmann.
ISBN: 0-672-22388-0 . . . . $19.95

☐ Basics of Audio and Visual Systems Design

Newcomers to the audio-visual industry, system designers, architects, contractors, equipment suppliers, students, teachers, consultants. . .all will find these NAVA-sanctioned fundamentals pertinent to system design procedures. Topics include image format, screen size and performance, front versus rear projection, projector output, audio, and the effective use of mirrors. Raymond Wadsworth.
ISBN: 0-672-22038-5 . . . . $15.95

☐ Handbook of Electronics Tables and Formulas (6th Edition)

This useful handbook contains all of the formulas and laws, constants and standards, symbols and codes, service and installation data, design data, and mathematical tables and formulas you would expect to find in this reference standard for the industry. The new edition contains computer programs for calculating many electrical and electronics formulas. Staff of Howard W. Sams.
ISBN: 0-672-22469-0 . . . . $19.95

☐ Modern Dictionary of Electronics (6th Edition)

This comprehensive dictionary clearly and accurately defines more than 23,000 technical terms dealing with computers, microelectronics, communications, semiconductors, and fiber optics. Over 3500 new entries and 5000 definitions, including abbreviations, cross-references, and acronyms, have been added, making this 6th edition the most up-to-date, all-inclusive electronics dictionary in the world. Rudolf E. Graf.
ISBN: 0-672-22041-5 . . . . $39.95

☐ CD-I and Interactive Videodisc Technology

This comprehensive reference guide explains how to program interactive videodiscs, CD-I, and digital interactive formats including CD-ROM and optical storage. It also discusses how the videodisc player and computer communicate, the pros and cons of authoring languages, interface cards, efficiency statistics, peripherals, types of networking systems, and surrogate computer controllers. Steve Lambert and Jane Sallis, Editors.
ISBN: 0-672-22513-1 . . . . $24.95

☐ John D. Lenk's Troubleshooting & Repair of Audio Equipment

This manual provides the most up-to-date data available and a simplified approach to practical troubleshooting and repair of major audio devices. Coverage includes dual cassette decks, compact disc players, lineartracking turntables, frequency-synthesized AM/FM tuners, IC amplifiers, and loudspeakers. John D. Lenk.
ISBN: 0-672-22517-4 . . . . $21.95

MORE
FROM
SAMS

☐ Sound System Engineering (2nd Edition)

This reference guide is written for the professional audio engineer. Everything from audio systems and loudspeaker directivity to sample design applications and specifications is covered in detail. Don Davis and Carolyn Davis.
ISBN: 0-672-21857-7 . . . .$39.95

☐ How to Read Schematics (4th Edition)

More than 100,000 copies in print! This update of a standard reference features expanded coverage of logic diagrams and a chapter on flowcharts. Beginning with a general discussion of electronic diagrams, the book systematically covers the various components that comprise a circuit. It explains logic symbols and their use in digital circuits, interprets sample schematics, analyzes the operation of a radio receiver, and explains the various kinds of logic gates. Review questions end each chapter. Donald E. Herrington.
ISBN 0-672-22457-7 . . . .$14.95

☐ Video Scrambling and Descrambling for Satellite and Cable TV

Learn the secrets of signal scrambling and descrambling (encoding and decoding) from the experts. The book discusses the theory and techniques needed to understand how over-the-air and cable signals are decoded and encoded. Projects are included such as building a scrambler and descrambler and how to build a video test generator with scrambling capability. Rudolf F. Graf and William Sheets.
ISBN: 0-672-22499-2 . . . .$19.95

☐ How To Build Speaker Enclosures

A practical guide to the whys and hows of constructing high-quality, top-performance speaker enclosures. A wooden box alone is not a speaker enclosure—size, baffling, sound insulation, speaker characteristics, and crossover points must all be carefully considered. Badmaieff and Davis.
ISBN 0-672-20520-3 . . . .$6.95

☐ The Howard W. Sams Crash Course in Digital Technology

Back by popular demand, the "crash course" format is applied to digital technology. This concise volume provides a solid foundation in digital fundamentals, state-of-the-art components, circuits, and techniques in the shortest possible time. It builds the specific knowledge and skills necessary to understand, build, test, and troubleshoot digital circuitry. No previous experience with digital is necessary. Louis E. Frenzel.
ISBN 0-672-21845-3 . . . .$19.95

☐ The Sams Hookup Book: Do-It-Yourself Connections for Your VCR

Here is all the information needed for simple to complex hook ups of home entertainment equipment. This step-by-step guide provides instructions to hook up a video cassette recorder to a TV, cable converter, satellite receiver, remote control, block converter, or video disk player. Howard W. Sams Engineering Staff.
ISBN: 0-672-22248-5 . . . .$4.95

☐ Video Cameras: Theory & Servicing

You can't fix it unless you know how it works. This entry-level technical primer on video camera servicing gives a clear, well-illustrated presentation of practical theory. From the image tube through the electronics to the final interface, all concepts are fully discussed. The final section on troubleshooting lets you put your new-found knowledge to work for profit. Gerald P. McGinty.
ISBN 0-672-22382-1 . . . .$18.95

☐ VCR Troubleshooting & Repair Guide

With approximately 25 million video cassette recorders on the market, this long-awaited book will help owners repair these popular machines when they break down. This helpful troubleshooting guide is for the electronics hobbyist, layperson, or technician who needs a preventive maintenance and troubleshooting reference for VCRs. Limited electronics experience is required to use it, but more sophisticated service and repair functions are discussed, and valuable information for the service technician is included. Robert C. Brenner.
ISBN: 0-672-22507-7 . . . .$19.95

☐ Video Production Guide

For those who want to learn how video production really works, this book contains real-world applications. Pre-production planning, creativity and organization, people handling, single- and multi-camera studio and on-location production, direction techniques, editing, special effects, and distribution of the finished production are addressed. This book is designed for working and aspiring producers/directors, broadcasters, schools, CATV personnel, and others in the industry. Lon McQuillin.
ISBN 0-672-22053-9 . . . .$28.95

☐ Video Tape Recorders (2nd Edition)

Helical tape recorders continue to be the equipment of choice in entertainment, industry, and broadcasting. This book shows you how to operate and service helical VTRs and includes numerous examples of recorder circuitry and mechanical transport systems. Harry Kybett.
ISBN 0-672-21521-7 . . . .$14.95

MORE
FROM
SAMS

☐ **Cable Television (2nd Edition)**

With this text, engineers and technicians can learn to examine each component in a cable system, alone and in relation to the system as a whole. Sections include component testing, troubleshooting, noise reduction, and system failure. An overview of fiber optics and communications satellites is also provided.
John Cunningham.
ISBN 0-672-21755-4 ....$15.95

☐ **The Home Satellite TV Installation and Troubleshooting Manual**

For the hobbyist or electronics buff, this title provides a comprehensive introduction to satellite communication theory, component operation, and the installation and troubleshooting of satellite systems. The authors detail the why's and wherefore's of selecting satellite equipment. If you are among the 100,000 people installing a satellite each month, this book is a must for your reference library. Baylin and Gale.
ISBN 0-672-22496-8 ....$29.95

☐ **Satellites Today**

Discusses key concepts of the fascinating universe of satellite communications in terms that laypersons can understand. Here is the history of satellite communications, the costs of satellite systems, system components, legal questions that have been and remain to be decided, and up-to-date coverage of the latest developments in satellites. Baylin and Gale.
ISBN 0-672-22492-5 ....$12.95

☐ **The Satellite TV Handbook**

Fed up with the high costs and limited access available with commercial cable TV? Learn how to legally and privately cut your cable TV costs in half, see TV shows that may be blacked out in your city, and pick up live and unedited network TV shows. Consider starting a mini-cable system in your apartment complex, or plug into video-supplied college courses, business news, and children's networks. It's all here! Shows how to buy or build and aim your own satellite antenna, and provides a list of programs available on satellites, channel by channel. Anthony T. Easton.
ISBN 0-672-22055-5 ....$16.95

☐ **The Hidden Signals on Satellite TV**

This is the authoritative guide for the technically oriented hobbyist. It details the satellite services available and demonstrates how to access and use the non-video transmissions available on your satellite receiver. The comprehensive coverage includes such non-video signals as audio channels, news services, teletext services, and commodity and stock market reports.
Harrington and Cooper.
ISBN 0-672-22491-7 ....$19.95
